T5-DHH-964

Holt Geometry

Chapter 10 Resource Book

HOLT, RINEHART AND WINSTON
A Harcourt Education Company
Orlando • Austin • New York • San Diego • London

Copyright © by Holt, Rinehart and Winston.

All rights reserved. No part of this publication may be reproduced or transmitted in any form or by any means, electronic or mechanical, including photocopy, recording, or any information storage and retrieval system, without permission in writing from the publisher.

Teachers using GEOMETRY may photocopy complete pages in sufficient quantities for classroom use only and not for resale.

HOLT and the **"Owl Design"** are trademarks licensed to Holt, Rinehart and Winston, registered in the United States of America and/or other jurisdictions.

Printed in the United States of America

If you have received these materials as examination copies free of charge, Holt, Rinehart and Winston retains title to the materials and they may not be resold. Resale of examination copies is strictly prohibited and is illegal.

Possession of this publication in print format does not entitle users to convert this publication, or any portion of it, into electronic format.

ISBN 0-03-042787-8

2 3 4 5 6 7 8 9 862 09 08 07 06

Contents

Blackline Masters

Copyright © by Holt, Rinehart and Winston.
All rights reserved.

Date ______________

Dear Family,

In this chapter, your child will learn about **three-dimensional figures** and **special relationships.** Your child will then learn to determine the surface areas and volumes of these three-dimensional figures.

You child will begin the lesson with a look at solid geometry.

There are four basic three-dimensional figures at which your child will be looking. These are outlined in the table below. Your child will learn more about these figures by using a **net.** A net is a diagram of a three-dimensional figure that can be folded to form the three-dimensional figure. The nets of these figures are also included in this table.

Shape	Description	Example	Net
Prism	Formed by two parallel congruent polygonal faces called *bases* connected by faces that are parallelograms	bases	
Cylinder	Formed by two parallel congruent circular bases and a curved surface that connects the bases	bases	
Pyramid	Formed by a polygonal base and triangular faces that meet at a common vertex.	vertex base	
Cone	Formed by a circular base and a curved surface that connects the base to a vertex.	vertex base	

Your child will also use formulas for these three-dimensional figures. One conjecture that your child will have to make is about the relationships between the vertices, edges, and faces of a polyhedron. A **polyhedron** is formed by four or more polygons that intersect only at their edges.

Euler's Formula says that for any polyhedron with V vertices, E edges, and F faces,

$$V - E + F = 2$$

Your child will then move on to determine the surface areas and volumes of different figures. It will be helpful to use the net of the figure to do this.

The surface area of a figure is different from the lateral area. The surface area is the total area of all faces and curved surfaces of a three-dimensional figure. The lateral area of a figure is the sum of the areas of the lateral faces. The lateral face is a face of a prism or pyramid that is not a base.

Copyright © by Holt, Rinehart and Winston.
All rights reserved.

The formulas needed to find these values are outlined below.

Shape	Lateral Area	Example	Surface Area	Example	Volume
Right Prism	$L = Ph$		$S = Ph + 2B$	h, a, b, c	$V = Bh$
Right Cylinder	$L = 2\pi rh$	r, h	$S = 2\pi rh + 2\pi r^2$	r, h, $2\pi r$	$V = \pi r^2 h$
Right Pyramid	$L = \frac{1}{2}P\ell$	ℓ	$S = \frac{1}{2}P\ell + B$	$P = 4s$, ℓ, s, s, s, s, s	$V = \frac{1}{3}Bh$
Right Cone	$L = \pi r\ell$	r, ℓ	$S = \pi r\ell + \pi r^2$	r, ℓ	$V = \frac{1}{3}\pi r^2 h$

Your child will also learn about the unique shape of spheres. A sphere is the locus of points in space that are a fixed distance from a given point called the center of a sphere. The radius of a sphere connects the center of the sphere to any point on the sphere.

The figure below shows the parts of a labeled sphere.

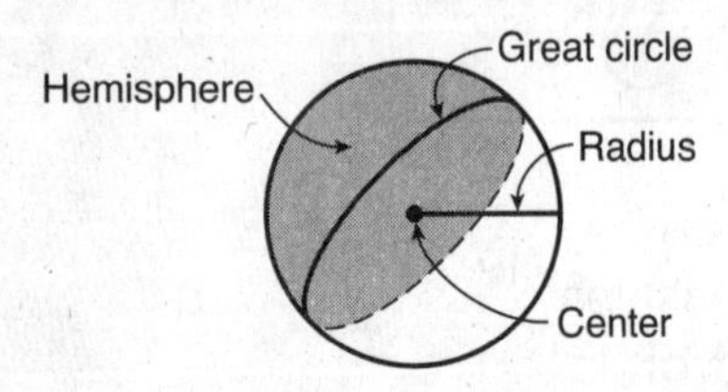

The volume of a sphere with radius r is $V = \frac{4}{3}\pi r^3$.

The surface area of a sphere with radius r is $S = 4\pi r^2$.

For additional resources, visit *go.hrw.com* and enter the keyword MG7 Parent.

Copyright © by Holt, Rinehart and Winston.
All rights reserved.

Name ______________________ Date ____________ Class ____________

LESSON 10-1

Practice A
Solid Geometry

For Exercises 1–4, match the given parts of the figure to the names.

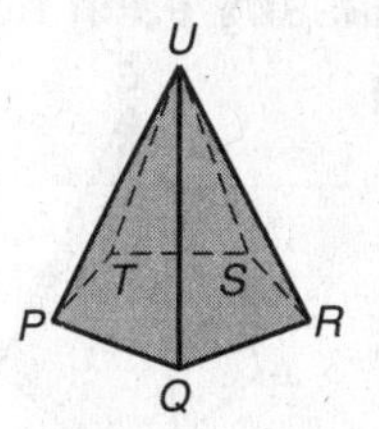

1. vertex ________
2. edge ________
3. face ________
4. base ________

a. triangle *PUT*
b. point *T*
c. pentagon *PQRST*
d. segment *PU*

Classify each figure. Name the vertices, edges, and bases.

5. 

Type of figure: ____________________

Vertices: ____________________

Edges: ____________________

____

Bases: ____________________

6. 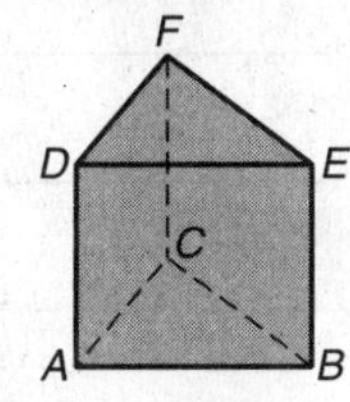

Type of figure: ____________________

Vertices: ____________________

Edges: ____________________

____

Bases: ____________________

Tell what kind of three-dimensional figure can be made from the given net.

7. 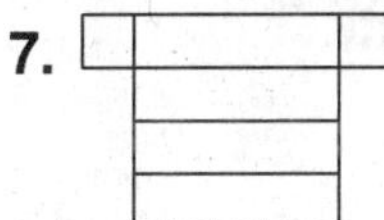

____

8. 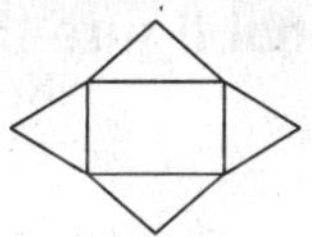

____

Tell what kind of shape each cross section makes.

9. 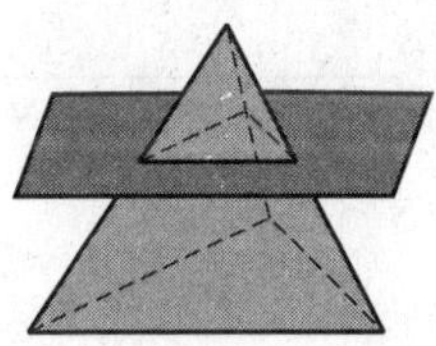

____

10. 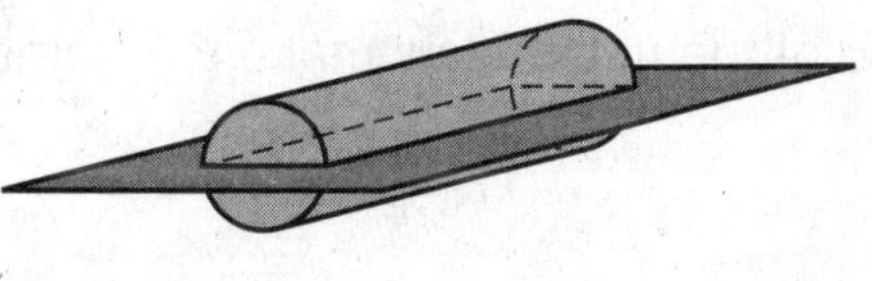

____

11. Soren cut several edges of a cereal box and then unfolded the box so it looks like this. Draw the box as it originally appeared and label the side lengths.

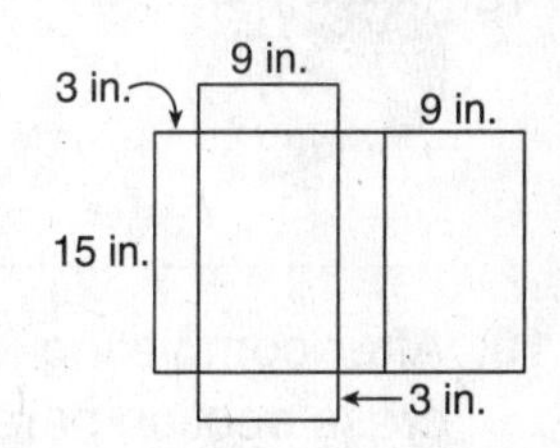

Copyright © by Holt, Rinehart and Winston.
All rights reserved.

Name ______________________ Date __________ Class __________

LESSON 10-1

Practice B
Solid Geometry

Classify each figure. Name the vertices, edges, and bases.

1. 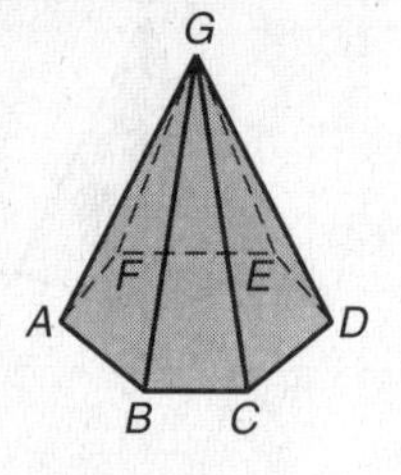

____

2. 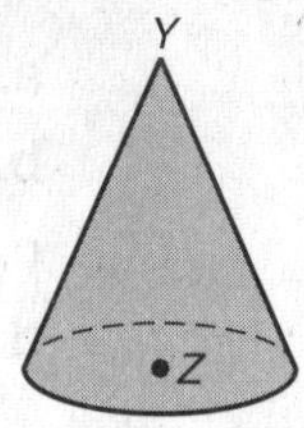

____

Name the type of solid each object is and sketch an example.

3. a shoe box

4. a can of tuna

Describe the three-dimensional figure that can be made from the given net.

5. 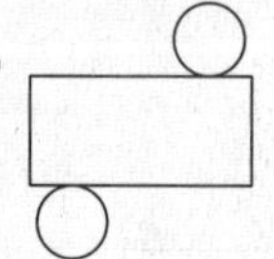

____

6. 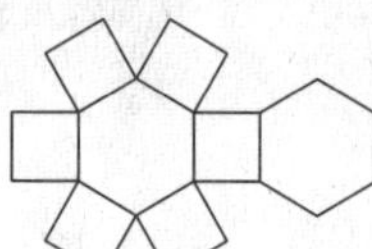

____

7. Two of the nets below make the same solid. Tell which one does not. ______________

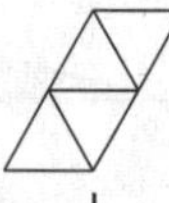
I

II

III

Describe each cross section.

8. 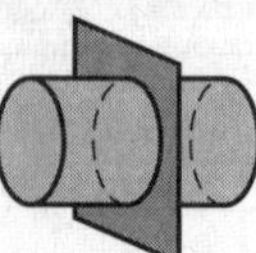

____

9. 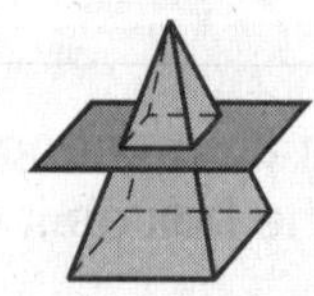

____

10. After completing Exercises 8 and 9, Lloyd makes a conjecture about the shape of any cross section parallel to the base of a solid. Write your own conjecture.

____

____

Copyright © by Holt, Rinehart and Winston.
All rights reserved.

Name ______________________ Date __________ Class __________

LESSON **10-1**

Practice C
Solid Geometry

A sphere is a three-dimensional figure bounded by all the points a fixed distance from a central point. Examples of a sphere include a globe and a basketball.

1. Name the two possible geometric figures that can result from the intersection of a plane and a sphere.

____

2. Tell whether a sphere has vertices, edges, faces, or bases. Name the two things that define a sphere.

____

____

A conic section is the intersection of a plane and a cone (or double cone). Many conic sections can be modeled by equations in x, y, x^2, and y^2. First graph each equation. Then sketch a plane and a cone so that their intersection has the same shape as the graph of the equation. (*Hint:* Sketch a double cone in Exercise 7.)

3. $y = x$

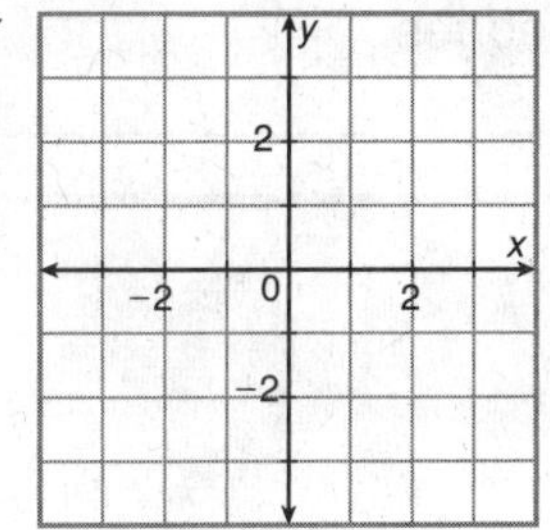

4. $y = x^2$

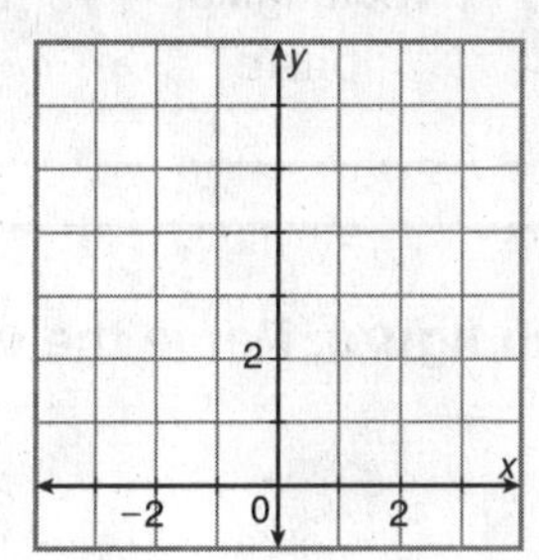

5. $x^2 + y^2 = 9$

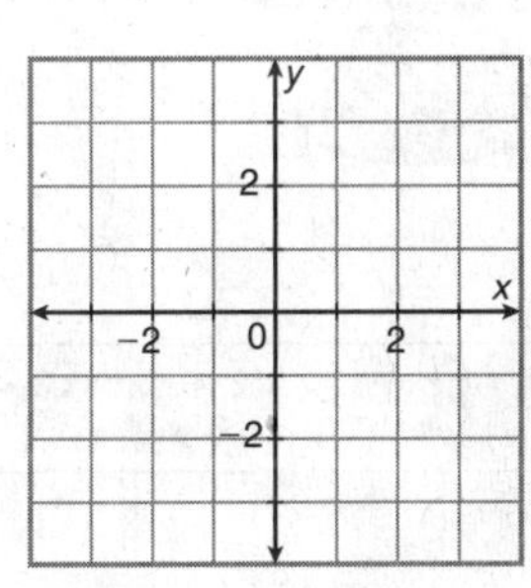

6. $\dfrac{x^2}{4} + \dfrac{y^2}{9} = 1$

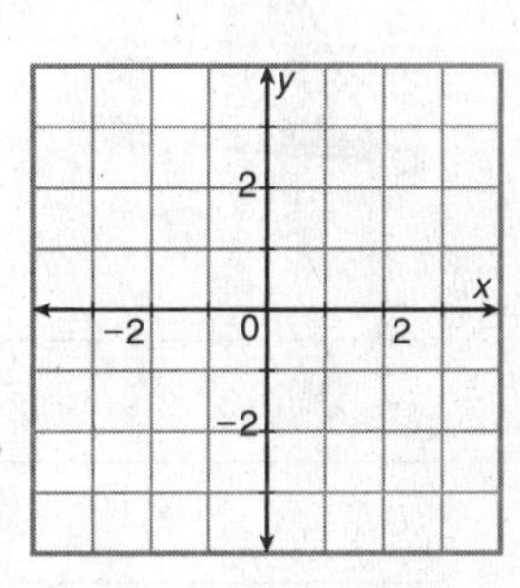

7. $y^2 - x^2 = 1$ (*Hint:* Remember that $y^2 = 1$ has two solutions.)

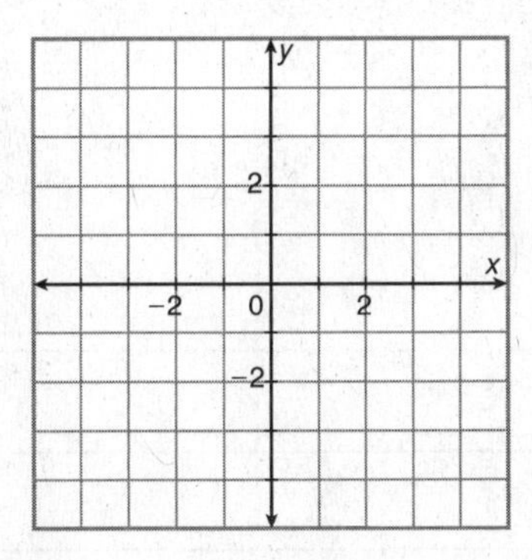

Copyright © by Holt, Rinehart and Winston.
All rights reserved.

Name ______________________ Date ____________ Class ____________

LESSON 10-1

Reteach

Solid Geometry

Three-dimensional figures, or *solids,* can have flat or curved surfaces.

Prisms and pyramids are named by the shapes of their *bases.*

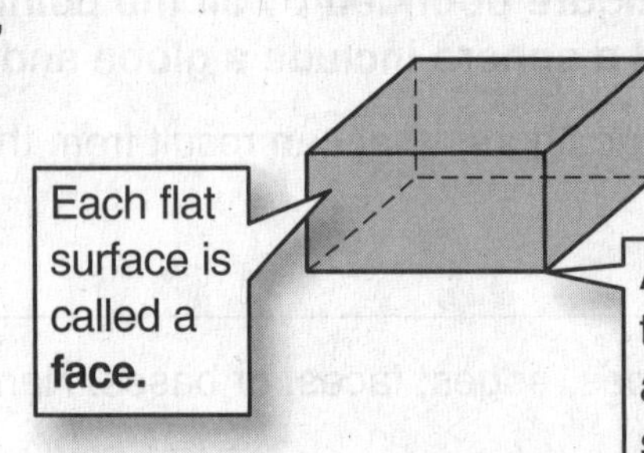

An **edge** is the segment where two faces intersect.
Each flat surface is called a **face.**
A **vertex** is the point where three or more faces intersect. In a cone, it is where the curved surface comes to a point.

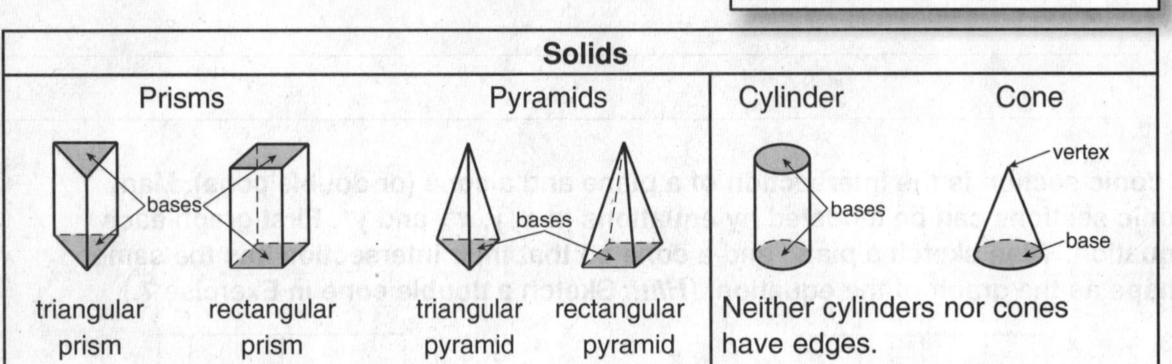

Solids
Prisms
Pyramids
Cylinder
Cone
bases
bases
bases
vertex
base
triangular prism
rectangular prism
triangular pyramid
rectangular pyramid
Neither cylinders nor cones have edges.

Classify each figure. Name the vertices, edges, and bases.

1. 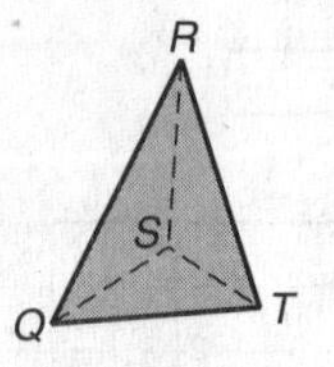

R
S
Q
T

____

____

____

2. 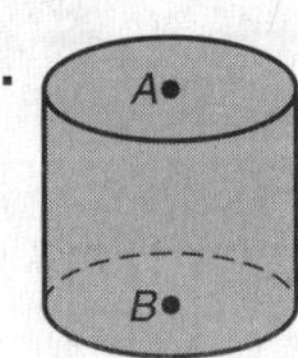

A
B

____

____

____

3. 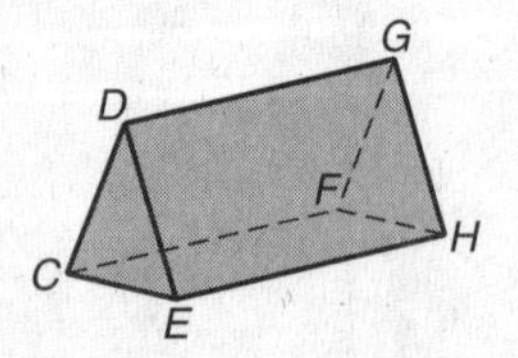

G
D
F
H
C
E

____

____

____

4. 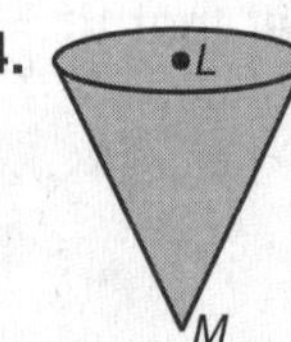

L
M

____

____

____

Copyright © by Holt, Rinehart and Winston.
All rights reserved.

Name ______________________ Date ____________ Class ____________

LESSON 10-1

Reteach

Solid Geometry continued

A **net** is a diagram of the surfaces of a three-dimensional figure. It can be folded to form the three-dimensional figure.

The net at right has one rectangular face. The remaining faces are triangles, and so the net forms a rectangular pyramid.

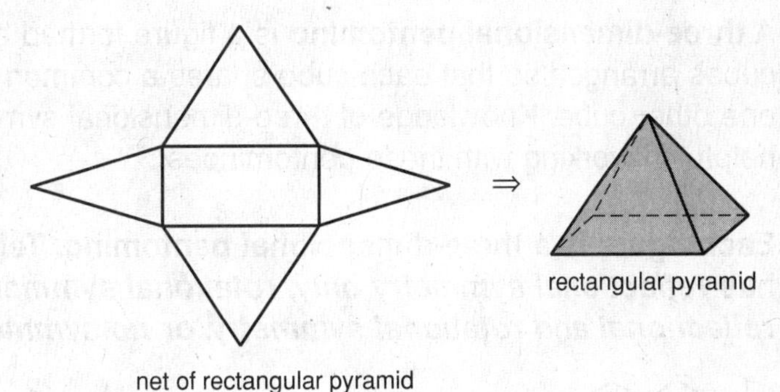

net of rectangular pyramid

A **cross section** is the intersection of a three-dimensional figure and a plane.

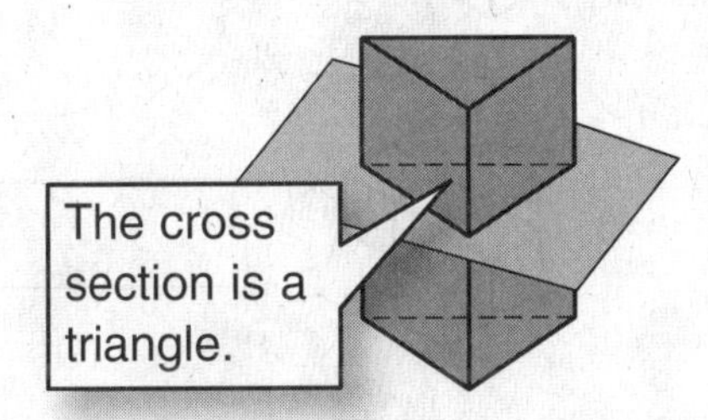

Describe the three-dimensional figure that can be made from the given net.

5.

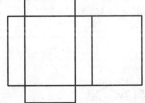

____

6.

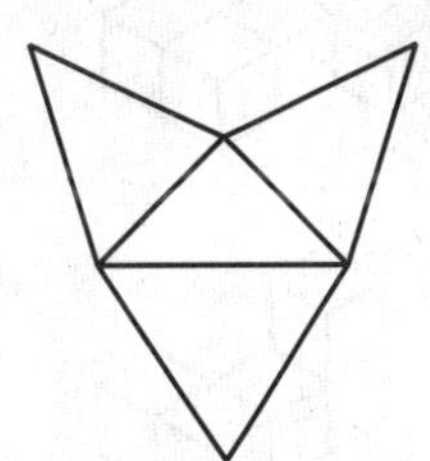

____

Describe each cross section.

7.

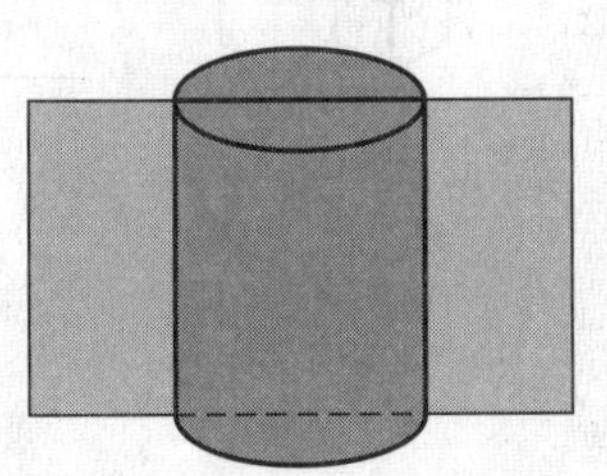

____

8.

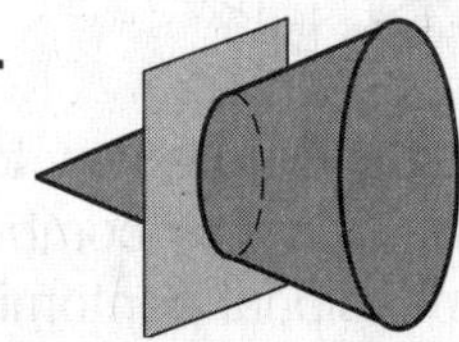

____

Copyright © by Holt, Rinehart and Winston.
All rights reserved.

Name ______________________ Date __________ Class __________

LESSON 10-1

Challenge

Three-Dimensional Pentominoes

A **three-dimensional pentomino** is a figure formed by five identical cubes arranged so that each cube shares a common face with at least one other cube. Knowledge of three-dimensional symmetries can be helpful in working with these pentominoes.

Each figure is a three-dimensional pentomino. Tell whether it has *reflectional symmetry only, rotational symmetry only, both reflectional and rotational symmetry,* or *no symmetry.*

1. 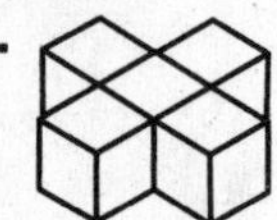 ______________________ ______________________

2.  ______________________ ______________________

3.  ______________________ ______________________

4.   ______________________ ______________________

In Exercises 5–7 a three-dimensional pentomino is shown at left. Name the pentomino A, B, or C to its right that is identical to it.

5. 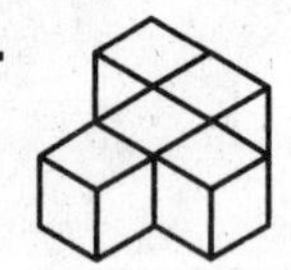 **A.** 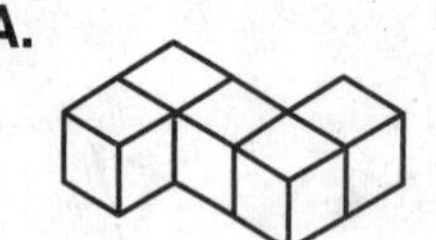 **B.** 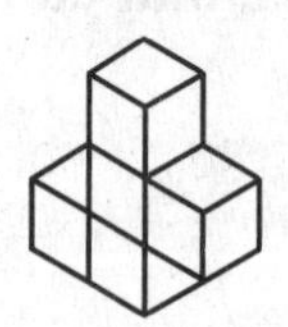 **C.** 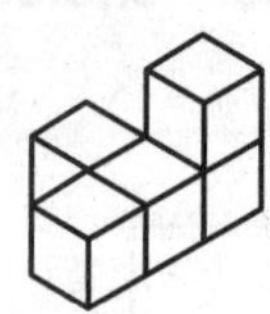 __________

6. 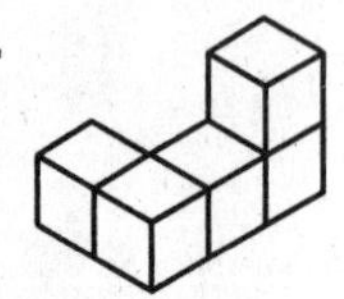 **A.** 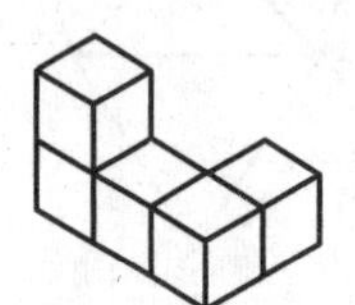 **B.** 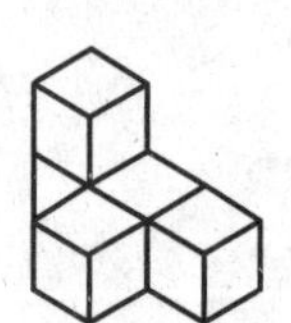 **C.** 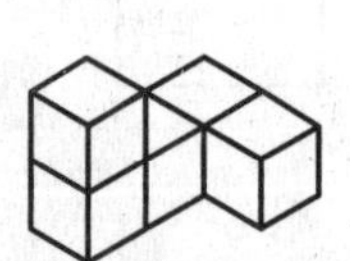 __________

7. 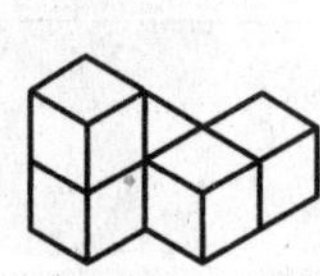 **A.** 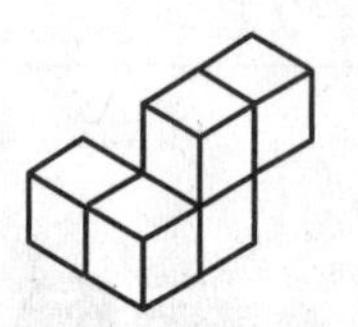 **B.** 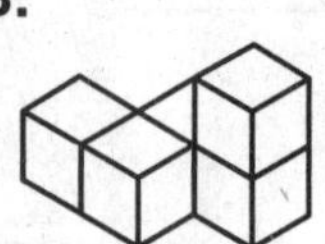 **C.**  __________

8. Create an original puzzle: Find a way to form a right rectangular prism by combining two or more three-dimensional pentominoes. Then make models of the pentominoes. Trade these "puzzle pieces" with a partner and see who can solve the other's puzzle first.

Copyright © by Holt, Rinehart and Winston.
All rights reserved.

Name ______________________ Date __________ Class __________

LESSON 10-1

Problem Solving

Solid Geometry

1. A slice of cheese is cut from the cylinder-shaped cheese as shown. Describe the cross section.

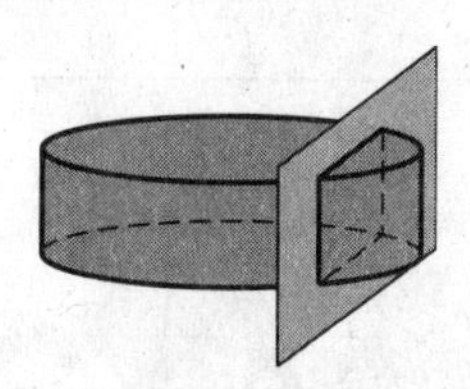

____

2. Mara has cut out five pieces of fabric to sew together to form a pillow. There are three rectangular pieces and two triangles. Describe the solid that will be formed.

____

3. A square pyramid is intersected by a plane as shown. Describe the cross section.

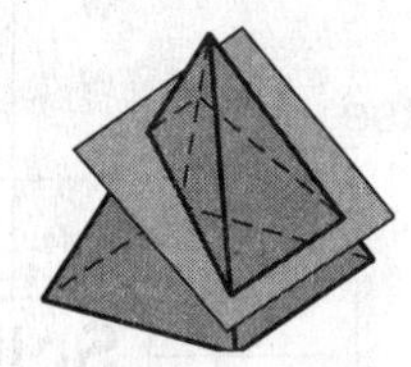

____

Choose the best answer.

4. A gift box is in the shape of a pentagonal prism. How many faces, edges, and vertices does the box have?

   A 6 faces, 10 edges, 6 vertices

   B 7 faces, 12 edges, 10 vertices

   C 7 faces, 15 edges, 10 vertices

   D 8 faces, 18 edges, 12 vertices

5. Which two solids have the same number of vertices?

   F rectangular prism and triangular pyramid

   G triangular prism and rectangular pyramid

   H rectangular prism and pentagonal pyramid

   J triangular prism and pentagonal pyramid

6. Which three-dimensional figure does the net represent?

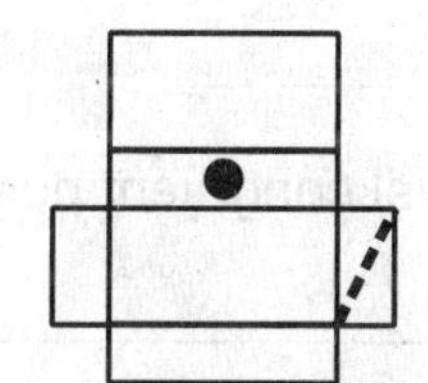

A

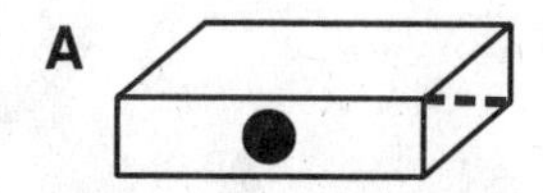

C

B

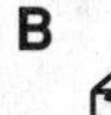

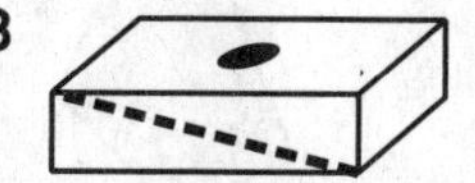

D

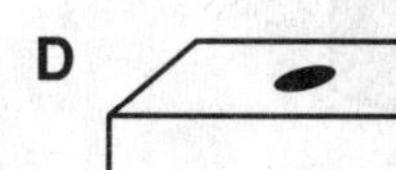

7. Which can be a true statement about the triangular prism whose net is shown?

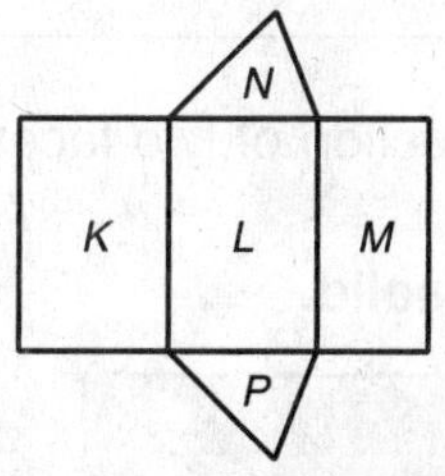

   F Faces *L* and *M* are perpendicular.

   G Faces *N* and *P* are perpendicular.

   H Faces *K* and *L* are parallel.

   J Faces *N* and *P* are parallel.

Copyright © by Holt, Rinehart and Winston.
All rights reserved.

Name ______________________ Date ____________ Class ____________

LESSON 10-1

Reading Strategies

Use a Graphic Aid

Solids are made up of flat surfaces and curved surfaces. A flat surface is called a **face.** The intersection of two faces is a segment called an **edge.** Three or more faces intersect at a **vertex.** Use the graphic aid below to understand solids.

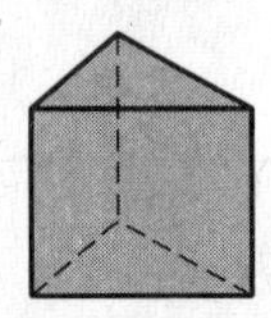

- Two parallel congruent polygonal bases
- The bases are connected by faces that are parallelograms.
- Named for the shape of its base

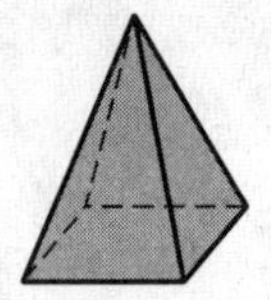

- One polygonal base
- The triangular faces meet at a common vertex.
- Named for the shape of its base

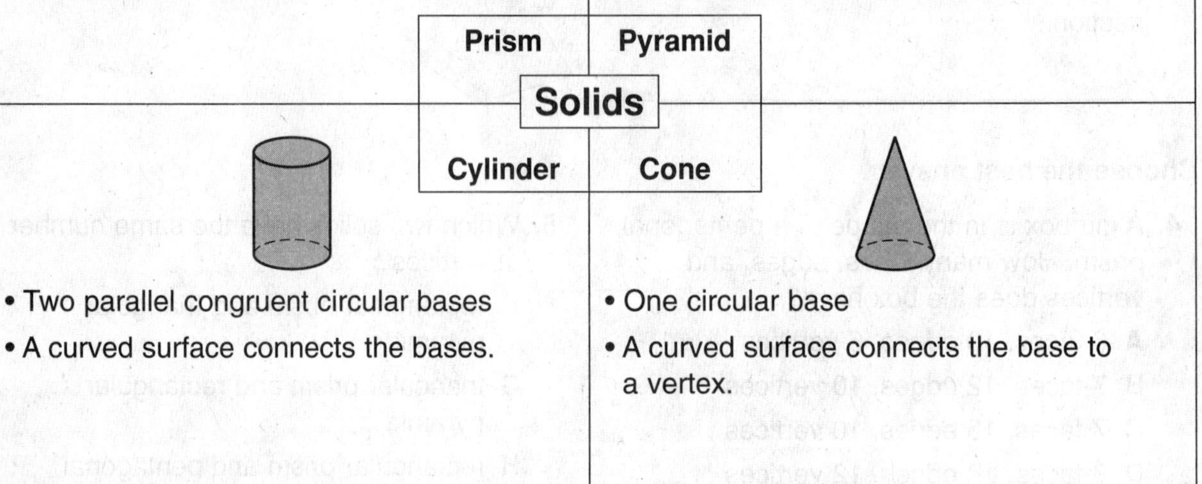

Prism
Pyramid
Solids
Cylinder
Cone
• Two parallel congruent circular bases
• A curved surface connects the bases.
• One circular base
• A curved surface connects the base to a vertex.

Answer the following.

1. Name two solids that have circular bases.

   ______________ ______________

2. A(n) ______________ has two parallel congruent polygonal bases.

3. The intersection of two faces is called a(n) ______________.

Identify each solid.

4. 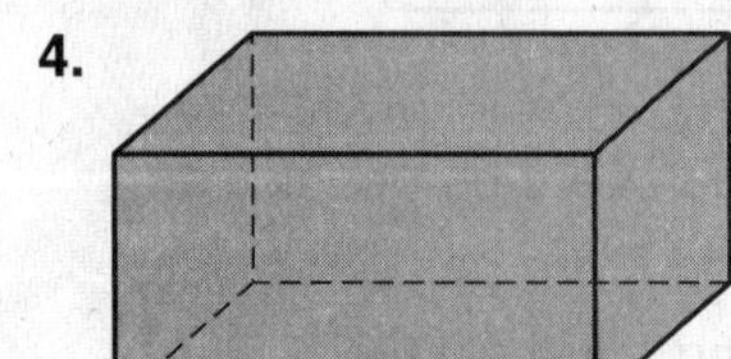

   ____

5. 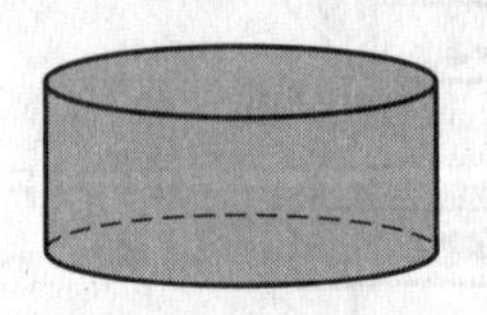

   ____

6. 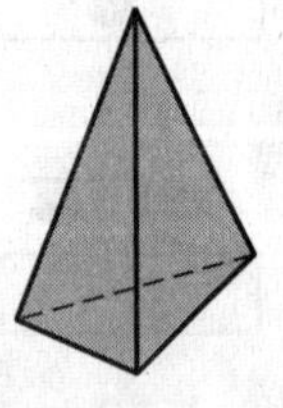

   ____

Copyright © by Holt, Rinehart and Winston.
All rights reserved.

Name ______________________ Date __________ Class __________

LESSON 10-2

Practice A

Representations of Three-Dimensional Figures

Draw all six orthographic views of each object (top, bottom, front, back, left, and right). Assume there are no hidden cubes. In your answers, use a dashed line to show that the edges touch and a solid line to show that the edges do not touch.

1. 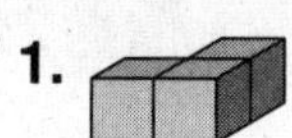

2. 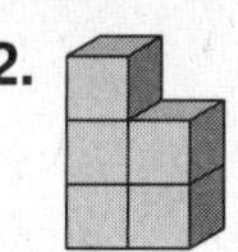

In an isometric drawing, every corner of a cube is on a dot in the grid.

3. Draw an isometric view of the object in Exercise 1.

4. Draw an isometric view of the object in Exercise 2.

5. Follow the steps to complete the drawing of a triangular prism in one-point perspective.

   a. Draw a dashed line from each vertex of the triangle to the vanishing point (point V).

   b. Use the dashed lines as guides to draw a triangle with sides parallel to the first triangle.

   c. Connect corresponding vertices of the two triangles. Use dashed lines for all hidden edges.

V

Determine whether each drawing represents the object at right. Assume there are no hidden cubes.

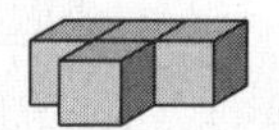

6. 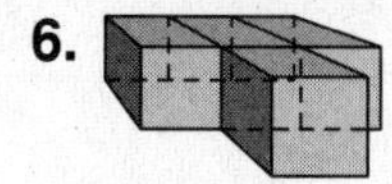

7. Top Bottom Left Right Front Back

______________________ ______________________

Copyright © by Holt, Rinehart and Winston. All rights reserved.

Name ____________________ Date ____________ Class ____________

LESSON 10-2

Practice B

Representations of Three-Dimensional Figures

Draw all six orthographic views of each object. Assume there are no hidden cubes. In your answers, use a dashed line to show that the edges touch and a solid line to show that the edges do not touch.

1.

2. 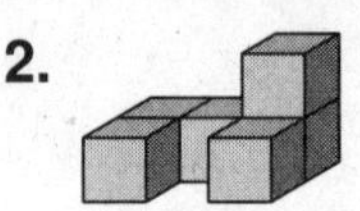

3. Draw an isometric view of the object in Exercise 1.

4. Draw an isometric view of the object in Exercise 2.

5. Draw a block letter T in one-point perspective.

6. Draw a block letter T in two-point perspective. (*Hint:* Draw the vertical line segment that will be closest to the viewer first.)

Determine whether each drawing represents the object at right. Assume there are no hidden cubes.

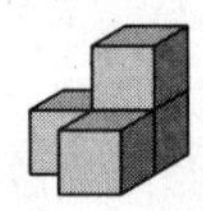

7. Top Bottom Left

Right Front Back

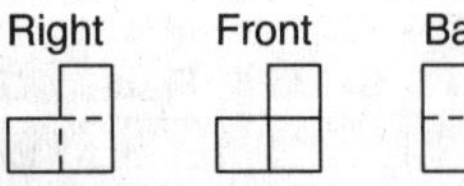

____

8. 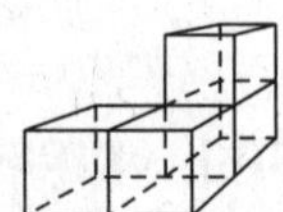

____

Copyright © by Holt, Rinehart and Winston.
All rights reserved.

Name ______________________ Date ___________ Class ___________

LESSON 10-2

Practice C

Representations of Three-Dimensional Figures

Draw an isometric view of each object based on the orthographic views provided.

1. Front Top

2. Top Right Left

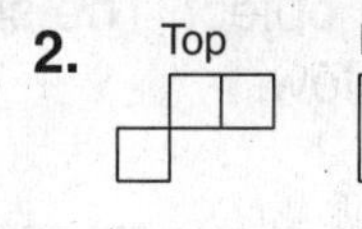

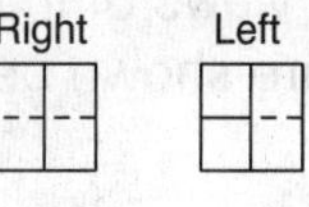

The object shown is made up of three pieces. Each piece is made of one or more adjoining cubes. Assume there are no hidden cubes.

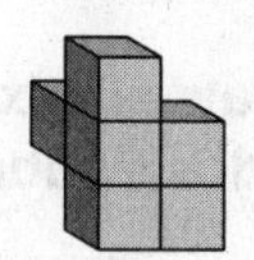

3. Assume each piece has a different shape and at least one piece is not a rectangular prism. Draw 3-D representations of the pieces.

4. Combine the three pieces you drew in Exercise 3 to make a rectangular prism. Draw the prism and shade the pieces so they can be distinguished.

5. Now suppose that two of the three pieces have the same shape. Draw the two same-shaped pieces. Then draw six possibilities for the third piece.

6. Four of the six possibilities you drew in Exercise 5 can form a 2-by-2-by-2 cube when joined together with another identical piece. Draw such a cube and shade the two pieces so they can be distinguished.

Copyright © by Holt, Rinehart and Winston.
All rights reserved.

Name ______________________ Date ____________ Class ____________

LESSON 10-2

Reteach

Representations of Three-Dimensional Figures

An **orthographic drawing** of a three-dimensional object shows six different views of the object. The six views of the figure at right are shown below.

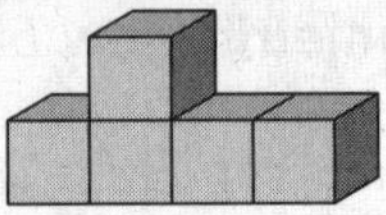

Top: 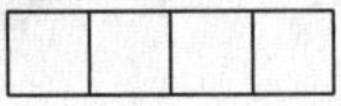 Bottom: 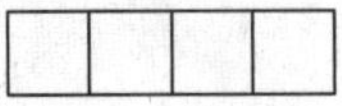 Front:

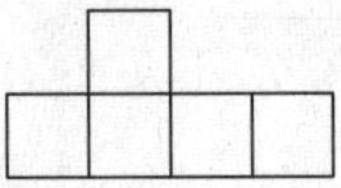

Back: 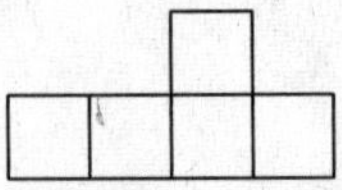 Left: 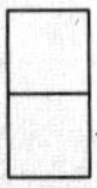 Right:

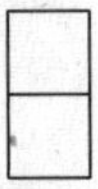

Draw all six orthographic views of each object. Assume there are no hidden cubes.

1.

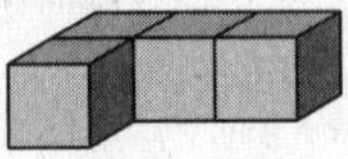

2. 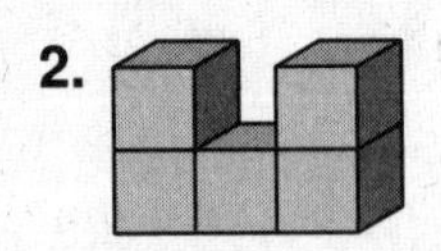

Copyright © by Holt, Rinehart and Winston.
All rights reserved.

Name ______________________ Date __________ Class __________

LESSON 10-2

Reteach

Representations of Three-Dimensional Figures continued

An **isometric drawing** is drawn on isometric dot paper and shows three sides of a figure from a corner view. A solid and an isometric drawing of the solid are shown.

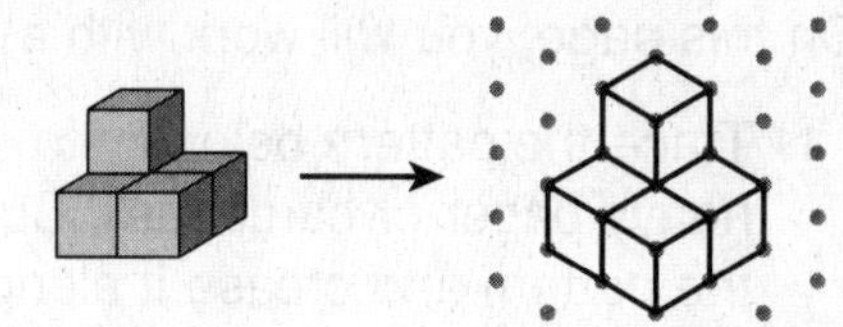

In a **one-point perspective drawing,** nonvertical lines are drawn so that they meet at a **vanishing point.** You can make a one-point perspective drawing of a triangular prism.

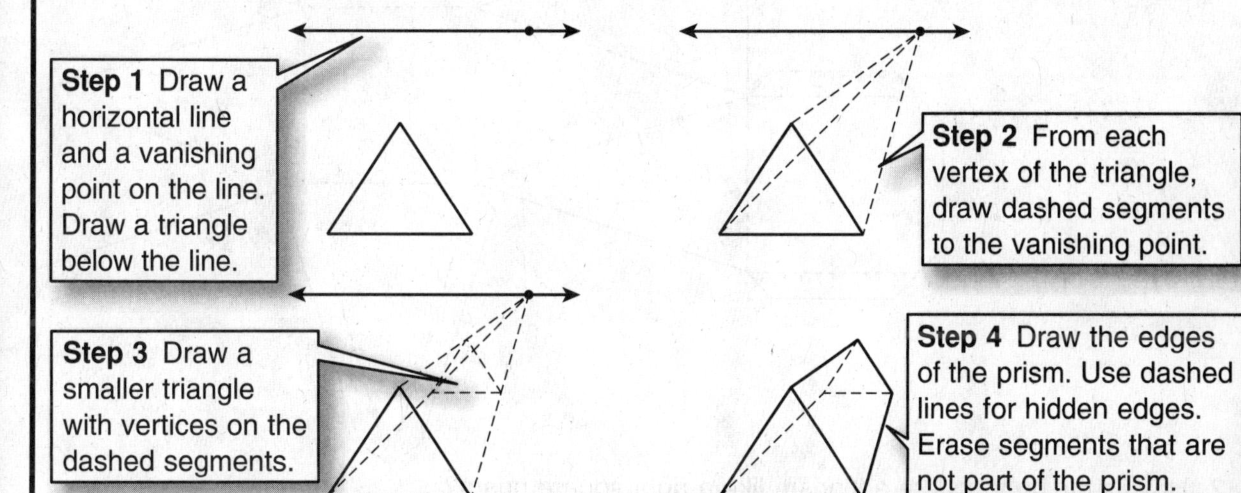

Draw an isometric view of each object. Assume there are no hidden cubes.

3.

4.

Draw each object in one-point perspective.

5. a triangular prism with bases that are obtuse triangles

6. a rectangular prism

Copyright © by Holt, Rinehart and Winston.
All rights reserved.

Name ______________________ Date ____________ Class ____________

LESSON 10-2

Challenge

Investigating Antiprisms

On this page, you will work with a type of polyhedron called an *antiprism.*

1. Trace the pattern below onto heavy paper or cardboard. Cut out the pattern and crease it along the dashed lines. Then use glue or tape to assemble it. The figure is a model of a *right square antiprism.*

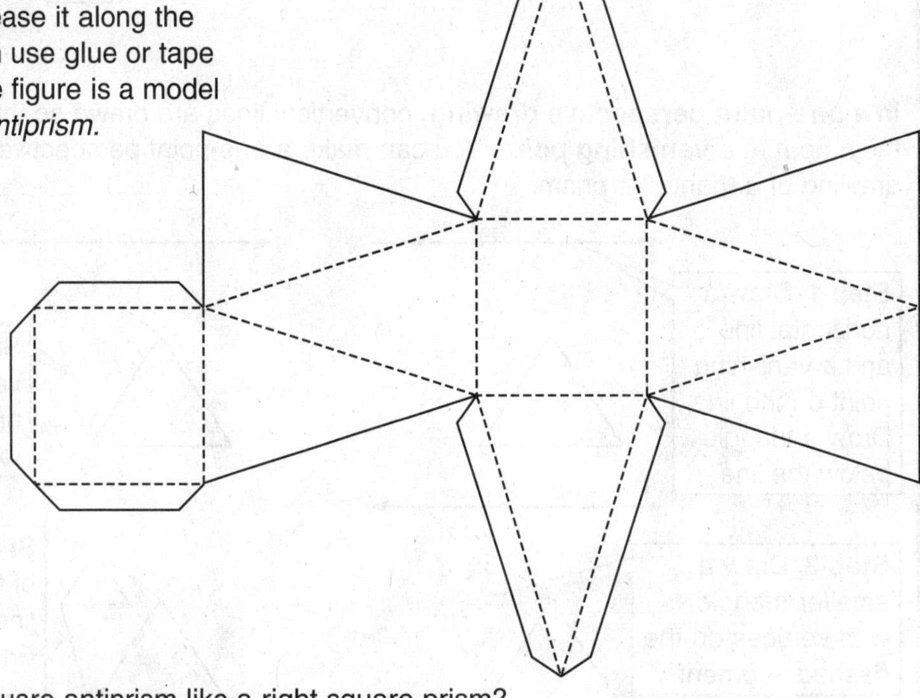

2. How is the right square antiprism like a right square prism? Name as many likenesses as you can.

____

____

____

3. How is the right square antiprism different from a right square prism? Name as many differences as you can.

____

____

4. On a separate sheet of paper, make a pattern for a right antiprism with two faces that are regular pentagons. Cut out and assemble the pattern. The figure is a *right regular pentagonal antiprism.*

Copyright © by Holt, Rinehart and Winston.
All rights reserved.

Name ______________________________ Date ____________ Class ____________

LESSON 10-2

Problem Solving

Representations of Three-Dimensional Figures

1. Describe the top, front, and side views of the figure.

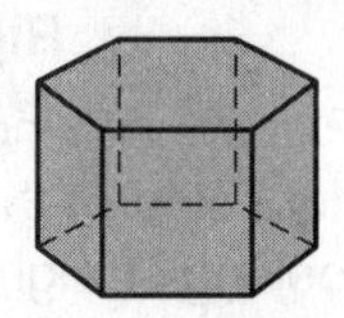

____

____

2. Erica used perspective to design the figure for a new logo. Describe the figure.

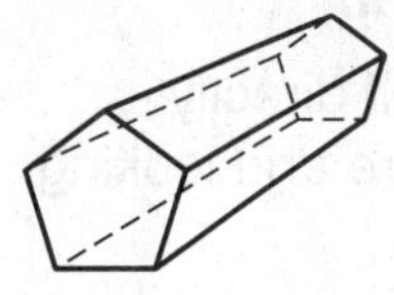

____

____

Choose the best answer.

3. Which is a true statement about the figure?

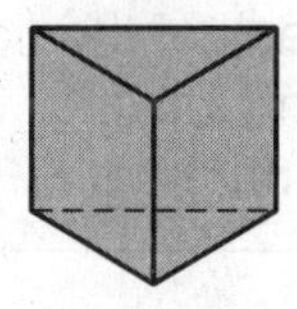

A The top view is a rectangle.

B A side view is a rectangle.

C A side view is a triangle.

D The front view is a triangle.

4. Which three-dimensional figure has these three views?

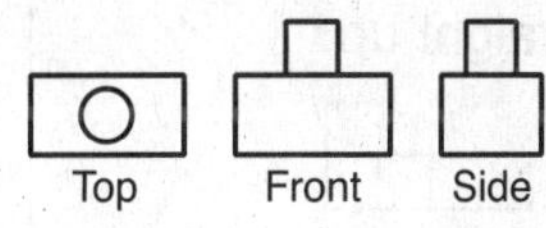

F 

H 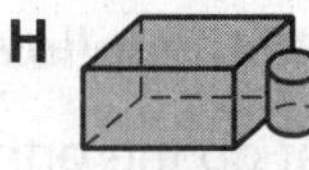

G 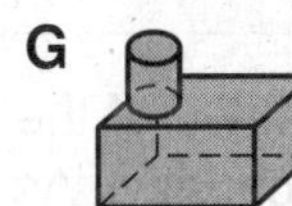

J 

5. Which drawing best represents the top view of the three-dimensional figure? Assume there are no hidden cubes.

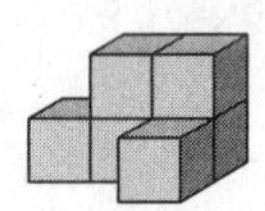

A 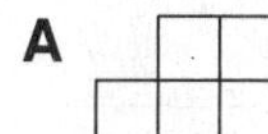

C 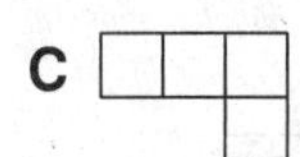

B 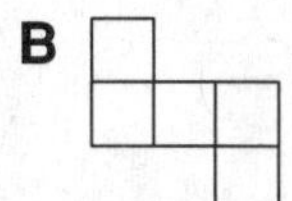

D 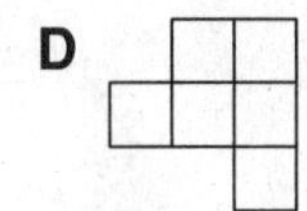

6. Which drawing best represents the side view of the building shown?

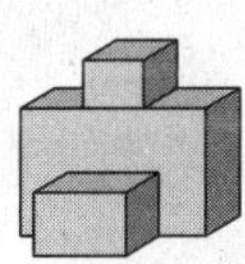

F

H 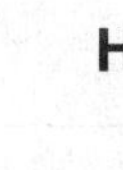

G 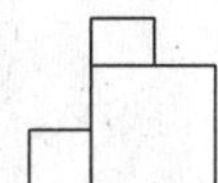

J 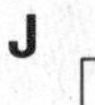

Copyright © by Holt, Rinehart and Winston.
All rights reserved.

Name ______________________ Date __________ Class __________

LESSON 10-2

Reading Strategies

Use a Concept Map

Orthographic views show three-dimensional objects from six different perspectives. Use the concept map to help you visualize orthographic views.

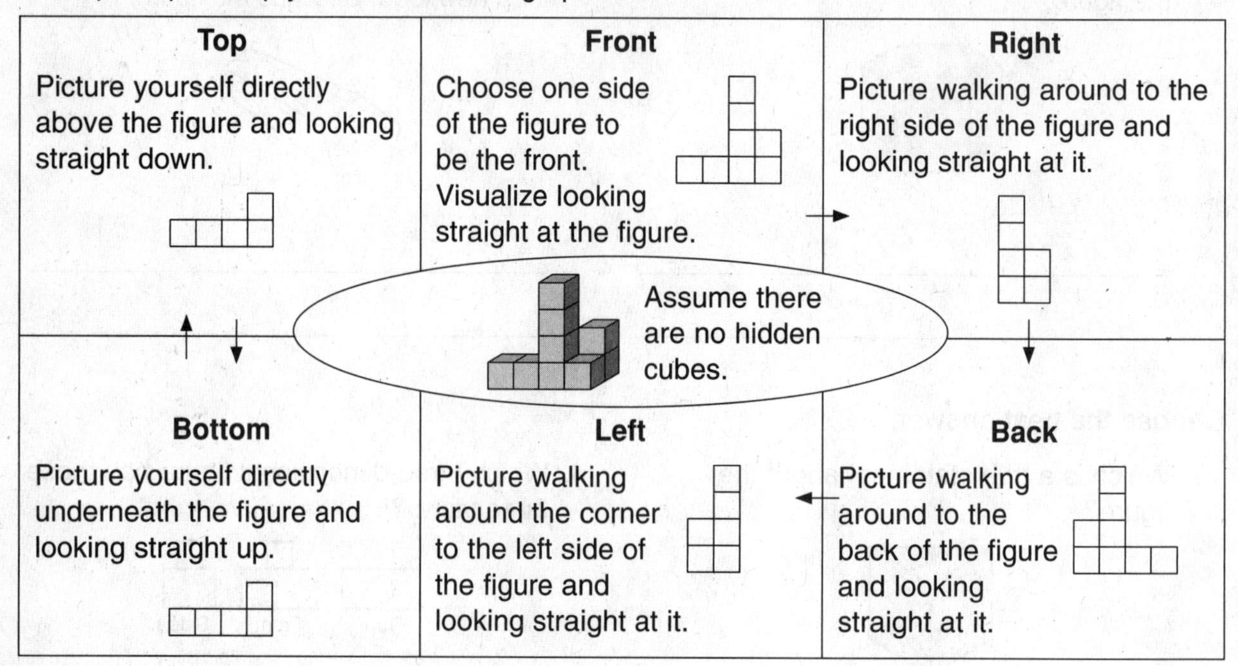

Complete the following.

1. What do the orthographic views of a three-dimensional object show?

____

2. Draw the six orthographic views of the object shown at right and label each view. Assume there are no hidden cubes.

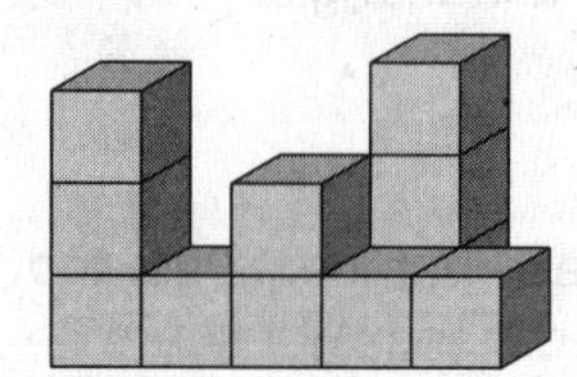

____________ ____________ ____________

____________ ____________ ____________

Copyright © by Holt, Rinehart and Winston.
All rights reserved.

Name ______________________ Date ____________ Class ____________

LESSON **10-3**

Practice A
Formulas in Three Dimensions

Match the letter of each formula to its name.

1. Euler's Formula ________
2. diagonal of a rectangular prism ________
3. distance in three dimensions ________
4. midpoint in three dimensions ________

a. $M\left(\frac{x_1 + x_2}{2}, \frac{y_1 + y_2}{2}, \frac{z_1 + z_2}{2}\right)$

b. $V - E + F = 2$

c. $d = \sqrt{\ell^2 + w^2 + h^2}$

d. $d = \sqrt{(x_2 - x_1)^2 + (y_2 - y_1)^2 + (z_2 - z_1)^2}$

Count the number of vertices, edges, and faces of each polyhedron. Use your results to verify Euler's Formula.

5. 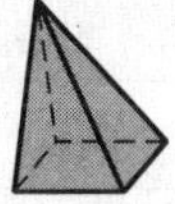

____

6.

____

For Exercises 7–9, use the formula for the length of a diagonal to find the unknown dimension in each polyhedron. Round to the nearest tenth.

7. the length of a diagonal of a cube with edge length 3 in. ________
8. the length of a diagonal of a 7-cm-by-10-cm-by-4-cm rectangular prism ________
9. the height of a rectangular prism with a 6-m-by-6-m base and a 9 m diagonal ________
10. A rectangular prism with length 3, width 2, and height 4 has one vertex at (0, 0, 0). Three other vertices are at (3, 0, 0), (0, 2, 0), and (0, 0, 4). Find the four other vertices. Then graph the figure.

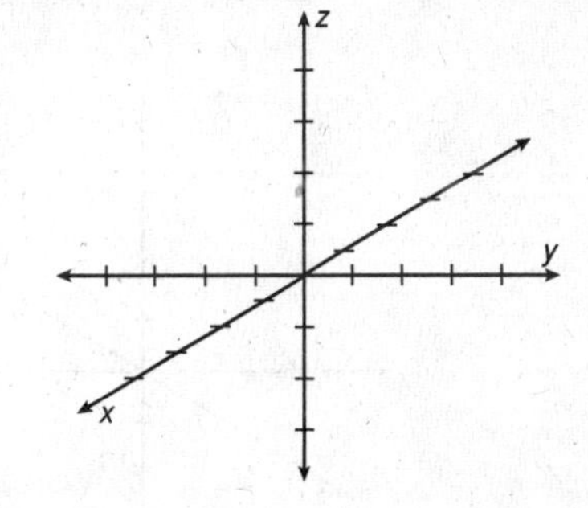

____

Use the formula for distance in three dimensions to find the distance between the given points. Use the midpoint formula in three dimensions to find the midpoint of the segment with the given endpoints. Round to the nearest tenth if necessary.

11. (0, 0, 0) and (2, 4, 6)

____

12. (1, 0, 5) and (0, 4, 0)

____

13. The world's largest ball of twine wound by a single individual weighs 17,400 pounds and has a 12-foot diameter. Roman climbs on top of the ball for a picture. To take the best picture, Lysandra moves 15 feet back and then 6 feet to her right. Find the distance from Lysandra to Roman. Round to the nearest tenth. ________

Copyright © by Holt, Rinehart and Winston.
All rights reserved.

Name ______________________ Date __________ Class __________

LESSON **10-3**

Practice B

Formulas in Three Dimensions

Find the number of vertices, edges, and faces of each polyhedron. Use your results to verify Euler's Formula.

1. 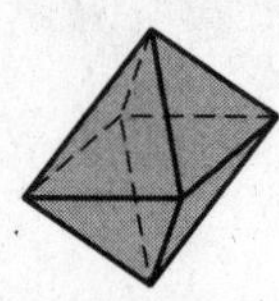

____

____

2. 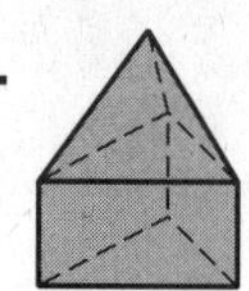

____

____

Find the unknown dimension in each polyhedron. Round to the nearest tenth.

3. the edge length of a cube with a diagonal of 9 ft __________

4. the length of a diagonal of a 15-mm-by-20-mm-by-8-mm rectangular prism __________

5. the length of a rectangular prism with width 2 in., height 18 in., and a 21-in. diagonal __________

Graph each figure.

6. a square prism with base edge length 4 units, height 2 units, and one vertex at (0, 0, 0)

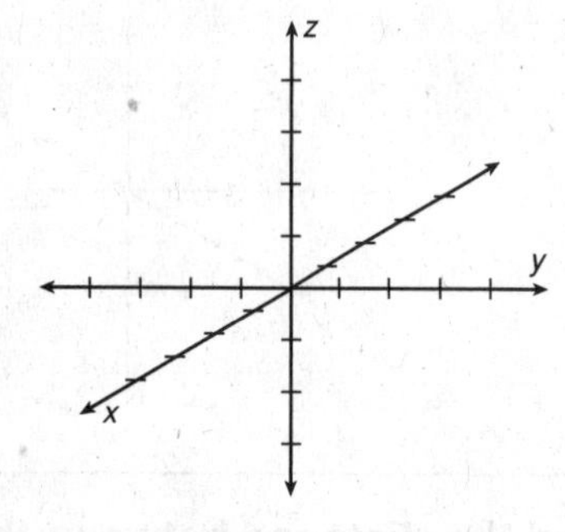

7. a cone with base diameter 6 units, height 3 units, and base centered at (0, 0, 0)

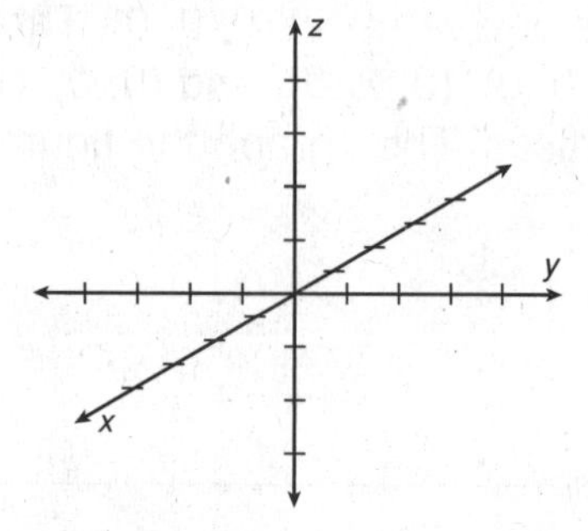

Find the distance between the given points. Find the midpoint of the segment with the given endpoints. Round to the nearest tenth if necessary.

8. (1, 10, 3) and (5, 5, 5)

____

9. (−8, 0, 11) and (2, −6, −17)

____

Copyright © by Holt, Rinehart and Winston.
All rights reserved.

Name ______________________ Date __________ Class __________

LESSON **10-3**

Practice C

Formulas in Three Dimensions

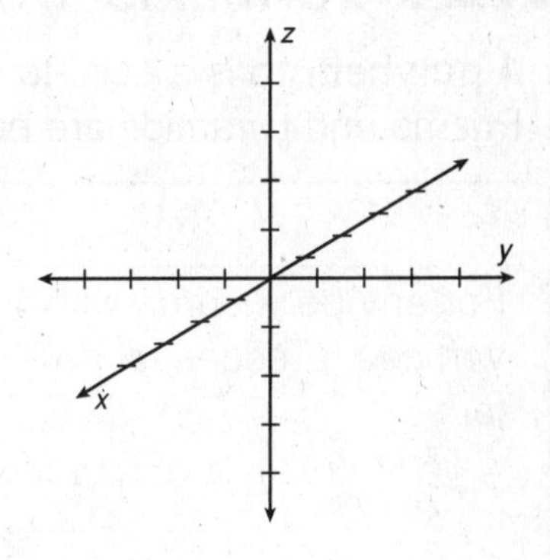

1. The distance from (0, 0, 0) to the surface of a solid is 4 units. Graph the solid.

2. Each edge of the solid shown in the figure measures 5 in. Find the length of $\overline{AB}$. Give an exact answer and an answer rounded to the nearest tenth.

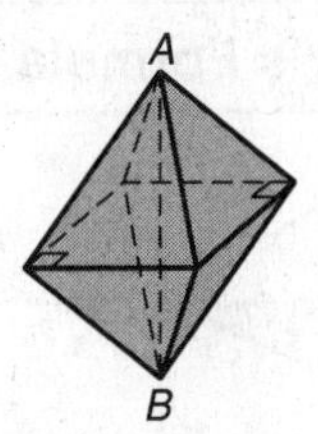

____

3. Find the length of $\overline{AB}$ if the bipyramid in Exercise 2 were based on a triangle rather than on a square. Round to the nearest tenth. _______________

4. Find the length of $\overline{AB}$ if the bipyramid in Exercise 2 were based on a pentagon rather than on a square. Round to the nearest tenth. _______________

5. If the bipyramid in Exercise 2 were based on a hexagon instead of a square, describe what sort of shape would result. Explain your answer.

6. The distance from $A(-2, 7, 0)$ to $B(3, 2, b)$ and from A to $C(3, 2, c)$ is 10 units. D lies on $\overline{BC}$ so that AD is the shortest distance from A to $\overline{BC}$. Find the coordinates of D without calculating. Explain how you got the answer.

7. A rectangular prism has vertices, in no particular order, at (−10, 8, 2), (−15, 8, 10), (−10, 5, 10), (−10, 5, 2), (−10, 8, 10), (−15, 5, 2), (−15, 5, 10), and (−15, 8, 2). Find the length of a diagonal of the prism. Round to the nearest tenth. _______________

8. Find the coordinates of a point that is equidistant from each of the eight vertices of the prism in Exercise 7. _______________

Tyrone has eight 1-in. cubes. He arranges all eight of them to make different rectangular prisms. Find the dimensions of the prisms based on the diagonal lengths given below.

9. $\sqrt{66}$ in. **10.** $\sqrt{21}$ in. **11.** $2\sqrt{3}$ in.

_______________ _______________ _______________

Copyright © by Holt, Rinehart and Winston.
All rights reserved.

Name ______________________ Date __________ Class __________

LESSON 10-3

Reteach

Formulas in Three Dimensions

A **polyhedron** is a solid formed by four or more polygons that intersect only at their edges. Prisms and pyramids are polyhedrons. Cylinders and cones are not.

Euler's Formula	
For any polyhedron with V vertices, E edges, and F faces, $V - E + F = 2$.	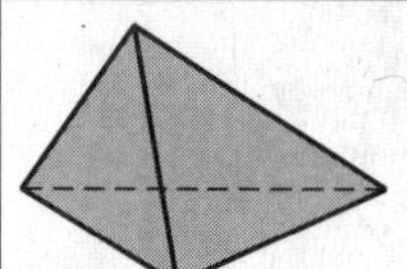 **Example** $V - E + F = 2$ Euler's Formula; $4 - 6 + 4 = 2$ $V = 4, E = 6, F = 4$; $2 = 2$. 4 vertices, 6 edges, 4 faces

Diagonal of a Right Rectangular Prism

The length of a diagonal d of a right rectangular prism with length ℓ, width w, and height h is

$$d = \sqrt{\ell^2 + w^2 + h^2}.$$

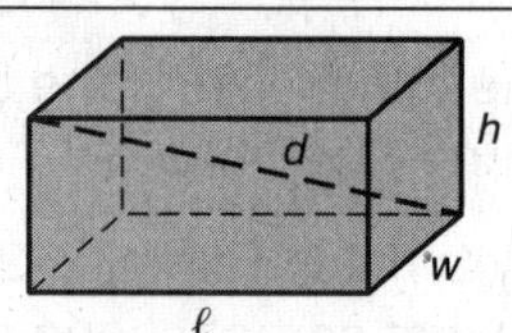

Find the height of a rectangular prism with a 4 cm by 3 cm base and a 7 cm diagonal.

$d = \sqrt{\ell^2 + w^2 + h^2}$	Formula for the diagonal of a right rectangular prism
$7 = \sqrt{4^2 + 3^2 + h^2}$	Substitute 7 for d, 4 for ℓ, and 3 for w.
$49 = 4^2 + 3^2 + h^2$	Square both sides of the equation.
$24 = h^2$	Simplify.
$4.9 \text{ cm} \approx h$	Take the square root of each side.

Find the number of vertices, edges, and faces of each polyhedron. Use your results to verify Euler's Formula.

1. 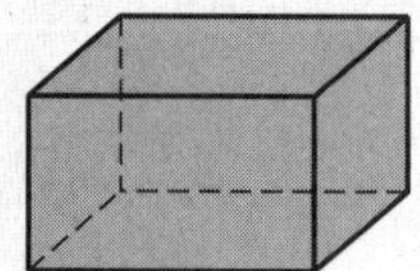

____

____

2. 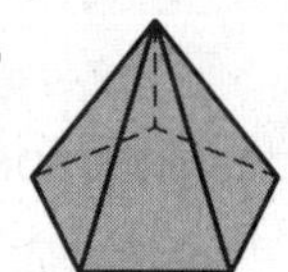

____

____

Find the unknown dimension in each figure. Round to the nearest tenth if necessary.

3. the length of the diagonal of a 6 cm by 8 cm by 11 cm rectangular prism

____

4. the height of a rectangular prism with a 4 in. by 5 in. base and a 9 in. diagonal

____

Copyright © by Holt, Rinehart and Winston.
All rights reserved.

Name ______________________ Date ____________ Class ____________

LESSON 10-3

Reteach

Formulas in Three Dimensions continued

A three-dimensional coordinate system has three perpendicular axes:

- *x*-axis
- *y*-axis
- *z*-axis

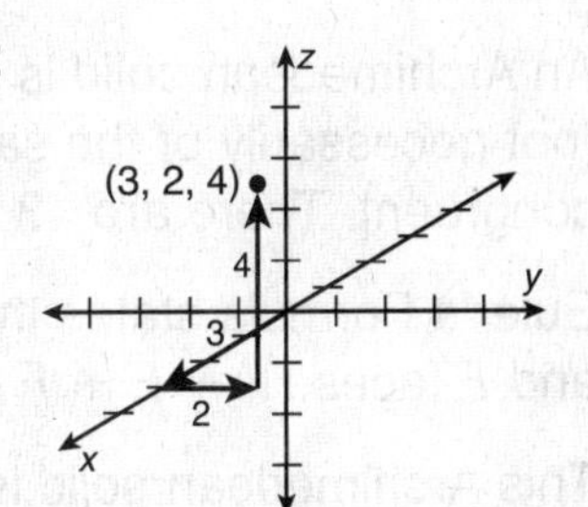

An *ordered triple* (*x*, *y*, *z*) is used to locate a point.
The point at (3, 2, 4) is graphed at right.

Formulas in Three Dimensions	
Distance Formula	The distance between the points (x_1, y_1, z_1) and (x_2, y_2, z_2) is $d = \sqrt{(x_2 - x_1)^2 + (y_2 - y_1)^2 + (z_2 - z_1)^2}$.
Midpoint Formula	The midpoint of the segment with endpoints (x_1, y_1, z_1) and (x_2, y_2, z_2) is $M\left(\frac{x_1 + x_2}{2}, \frac{y_1 + y_2}{2}, \frac{z_1 + z_2}{2}\right)$.

Find the distance between the points (4, 0, 1) and (2, 3, 0). Find the midpoint of the segment with the given endpoints.

$d = \sqrt{(x_2 - x_1)^2 + (y_2 - y_1)^2 + (z_2 - z_1)^2}$ Distance Formula

$= \sqrt{(2 - 4)^2 + (3 - 0)^2 + (0 - 1)^2}$ $(x_1, y_1, z_1) = (4, 0, 1), (x_2, y_2, z_2) = (2, 3, 0)$

$= \sqrt{4 + 9 + 1}$ Simplify.

$= \sqrt{14} \approx 3.7$ units Simplify.

The distance between the points (4, 0, 1) and (2, 3, 0) is about 3.7 units.

$M\left(\frac{x_1 + x_2}{2}, \frac{y_1 + y_2}{2}, \frac{z_1 + z_2}{2}\right) = M\left(\frac{4 + 2}{2}, \frac{0 + 3}{2}, \frac{1 + 0}{2}\right)$ Midpoint Formula

$= M(3, 1.5, 0.5)$ Simplify.

The midpoint of the segment with endpoints (4, 0, 1) and (2, 3, 0) is *M*(3, 1.5, 0.5).

Find the distance between the given points. Find the midpoint of the segment with the given endpoints. Round to the nearest tenth if necessary.

5. (0, 0, 0) and (6, 8, 2)

____

6. (0, 6, 0) and (4, 8, 0)

____

7. (9, 1, 4) and (7, 0, 7)

____

8. (2, 4, 1) and (3, 3, 5)

____

Copyright © by Holt, Rinehart and Winston.
All rights reserved.

Name ________________________________ Date ____________ Class ____________

LESSON 10-3

Challenge

Formulas in Three Dimensions

An Archimedean solid is a polyhedron whose faces are regular polygons (not necessarily of the same type) and whose polyhedral angles are all congruent. There are 13 such solids, of which only 5 are regular.

Euler's Formula states that for any polyhedron with V vertices, E edges, and F faces, $V - E + F = 2$.

This Archimedean solid is called the Great Rhombicosidodecahedron. The two-dimensional drawing is its net. A Great Rhombicosidodecahedron has 120 vertices and 180 edges.

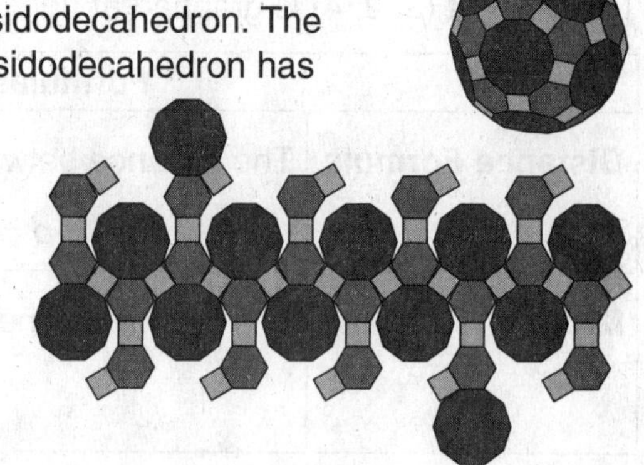

The notation for the two-dimensional figures that form the faces of a polyhedron is f_3 for triangular faces, f_4 for quadrilateral faces, f_5 for pentagonal faces, and so on. The Great Rhombicosidodecahedron has 30 quadrilateral faces ($f_4 = 30$), 20 hexagonal faces ($f_6 = 20$), and 12 decagonal faces ($f_{10} = 12$).

1. How many faces does the Great Rhombicosidodecahedron have? ____________

Use the figure for Exercise 2. This Archimedean solid is called a Snub Dodecahedron. It has 150 edges and 92 faces. The faces are as follows: $f_3 = 80$ and $f_5 = 12$.

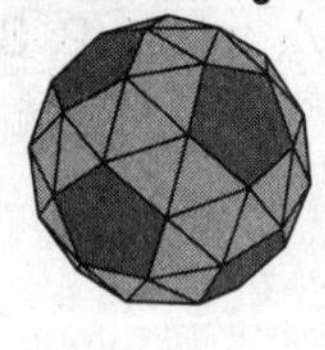

2. How many vertices does the Snub Dodecahedron have? ____________

Use the figure for Exercises 3–6. This Archimedean solid is called a Truncated Tetrahedron.

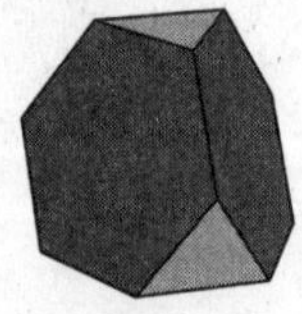

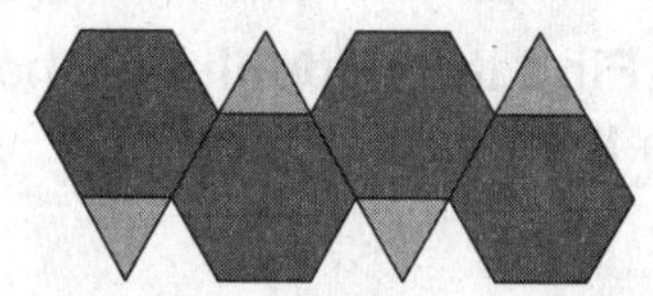

3. How many faces does the Truncated Tetrahedron have? ____________

4. How many edges does the Truncated Tetrahedron have? (*Hint:* Count all the sides of all the faces and divide by 2. Each edge consists of two sides touching.) ____________

5. How many vertices does the Truncated Tetrahedron have? ____________

6. Using proper notation, list the types of faces that are on a Truncated Tetrahedron and the number of each type. ____________

Copyright © by Holt, Rinehart and Winston.
All rights reserved.

Name ______________________ Date __________ Class __________

LESSON 10-3

Problem Solving

Formulas in Three Dimensions

1. What is the height of the rectangular prism? Round to the nearest tenth if necessary.

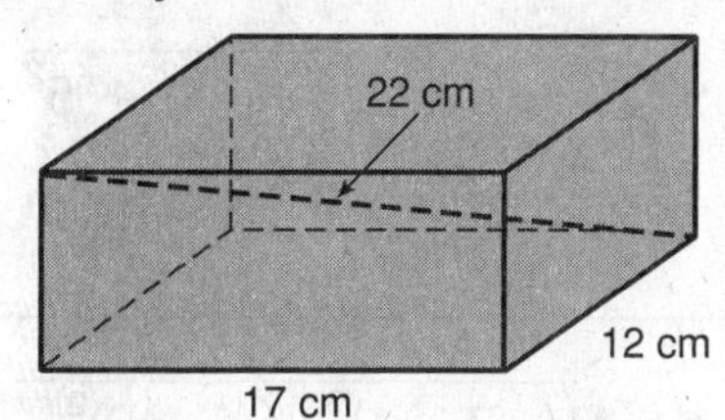

____

2. After lunch, Justin leaves the cafeteria to go to class, which is 22 feet north and 15 feet west of where he ate. The classroom is on the second floor, so it is 10 feet above the cafeteria. What is the actual distance between where Justin ate lunch and the classroom? Round to the nearest tenth.

____

3. Emily's hotel room is 18 feet south and 40 feet west of the pool. Her cousin Amber's hotel room is 22 feet north, 45 feet east, and 20 feet up on the third floor. How far apart are Emily's and Amber's rooms? Round to the nearest tenth.

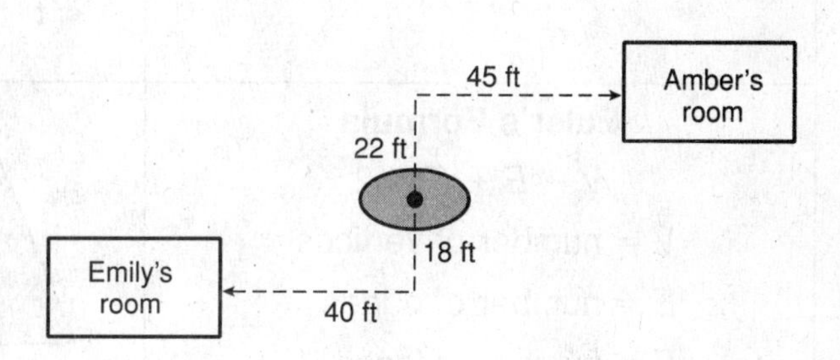

____

Choose the best answer.

4. How many faces, edges, and vertices does an octagonal pyramid have?
   - **A** 7 faces, 12 edges, 7 vertices
   - **B** 9 faces, 12 edges, 8 vertices
   - **C** 9 faces, 16 edges, 9 vertices
   - **D** 10 faces, 24 edges, 16 vertices

5. Which does NOT describe a polyhedron?
   - **F** 8 vertices, 12 edges, 6 faces
   - **G** 8 vertices, 10 edges, 6 faces
   - **H** 6 vertices, 9 edges, 5 faces
   - **J** 6 vertices, 10 edges, 6 faces

6. Point R has coordinates (8, 6, 1), and the midpoint of $\overline{RS}$ is $M(15, -2, 7)$. Which is the best estimate for the distance between point R and point S?
   - **A** 10.0 units
   - **B** 12.2 units
   - **C** 21.0 units
   - **D** 24.4 units

7. A rectangular prism has the following vertices. What is the volume of the prism?

   $A(0, 0, 4)$ $B(-4, 0, 0)$
   $C(-4, 2, 0)$ $D(0, 2, 0)$
   $E(0, 0, 0)$ $F(-4, 0, 4)$
   $G(-4, 2, 4)$ $H(0, 2, 4)$

   - **F** 4 units^3
   - **G** 16 units^3
   - **H** 32 units^3
   - **J** 64 units^3

Copyright © by Holt, Rinehart and Winston.
All rights reserved.

Name ______________________ Date ____________ Class ____________

LESSON 10-3

Reading Strategies

Use a Table

The table below shows some of the formulas used in three dimensions.

Formula	Diagram	Example
Length of Diagonal of a Right Rectangular Prism $d = \sqrt{\ell^2 + w^2 + h^2}$	5 cm, 4 cm, 6 cm, d	$d = \sqrt{6^2 + 4^2 + 5^2}$ $d = \sqrt{77} \approx 8.8$ cm
Distance Formula $d = \sqrt{(x_2 - x_1)^2 + (y_2 - y_1)^2 + (z_2 - z_1)^2}$	z, y, x, (1, 5, 6), d, (−7, 5, 0), 4	$d = \sqrt{(-7 - 1)^2 + (5 - 5)^2 + (0 - 6)^2}$ $d = \sqrt{(-8)^2 + (0)^2 + (-6)^2}$ $d = \sqrt{100} = 10$ units
Euler's Formula $V - E + F = 2$ V = number of vertices E = number of edges F = number of faces		Vertices: 5 Edges: 8 Faces: 5 $5 - 8 + 5 = 2$
Midpoint Formula $M = \left(\frac{x_1 + x_2}{2}, \frac{y_1 + y_2}{2}, \frac{z_1 + z_2}{2}\right)$	z, M, (6, −5, 10), (−2, 3, 8), 4, y, x	$\left(\frac{-2 + 6}{2}, \frac{3 + (-5)}{2}, \frac{8 + 10}{2}\right)$ $M = (2, -1, 9)$

Answer the following.

1. Write Euler's Formula in words.

____

____

____

2. Find the length of the diagonal of a 3 centimeter by 4 centimeter by 10 centimeter rectangular prism. Round to the nearest tenth.

____

Find the distance between the given points. Find the midpoint of the segment with the given points as endpoints. Round to the nearest tenth if necessary.

3. (2, 4, 5) and (6, 3, 1)

$d =$ ______________________

$M =$ ______________________

4. (−1, 4, 7) and (5, 0, −5)

$d =$ ______________________

$M =$ ______________________

Copyright © by Holt, Rinehart and Winston.
All rights reserved.

Name ________________________________ Date ______________ Class ______________

LESSON 10-4

Practice A

Surface Area of Prisms and Cylinders

Write each formula.

1. lateral area of a right prism with base perimeter *P* and height *h* ______________
2. lateral area of a right cylinder with radius *r* and height *h* ______________
3. surface area of a right prism with lateral area *L* and base area *B* ______________
4. surface area of a cube with edge length *s* ______________
5. surface area of a right cylinder with radius *r* and height *h* ______________

Find the lateral area and surface area of each right prism.

6. 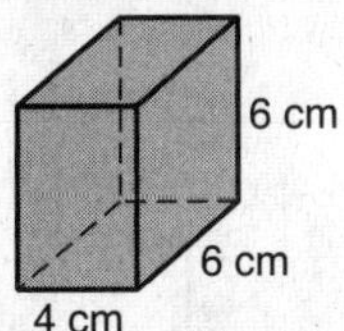

the rectangular prism

____

7. 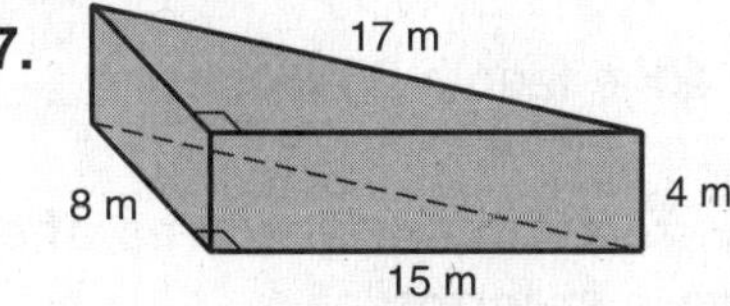

the triangular prism

____

8. a cube with edge length 2 ft ______________________

Find the lateral area and surface area of each right cylinder. Give your answers in terms of π.

9. 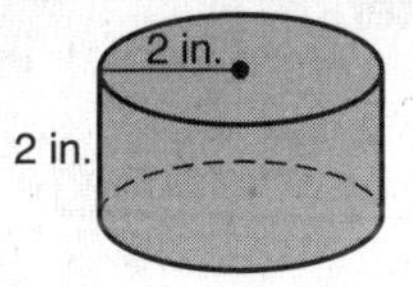

____

10. a cylinder with a radius of 3 mm and a height of 10 mm

____

A builder drills a hole through a cube of concrete, as shown in the figure. This cube will be an outlet for a water tap on the side of a house. Complete Exercises 11–14 to find the surface area of the figure. Round to the nearest tenth if necessary.

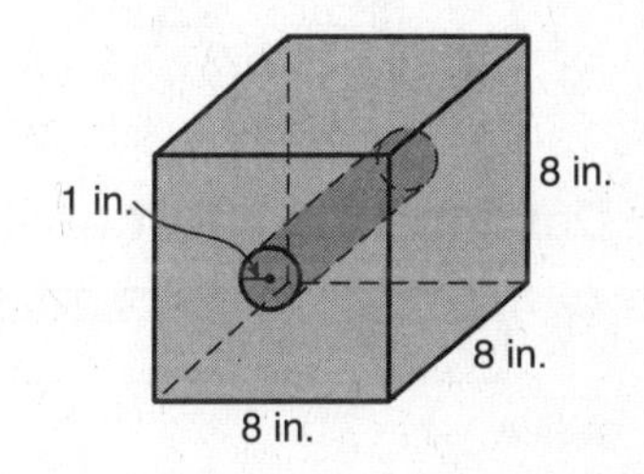

11. Find the surface area of the cube. ______________
12. Find the lateral area of the cylinder. ______________
13. Find twice the base area of the cylinder. ______________
14. The surface area of the figure is the surface area of the prism plus the lateral area of the cylinder minus twice the base area of the cylinder. Find the surface area of the figure. ______________

Copyright © by Holt, Rinehart and Winston.
All rights reserved.

Name ______________________ Date ____________ Class ____________

LESSON 10-4

Practice B

Surface Area of Prisms and Cylinders

Find the lateral area and surface area of each right prism. Round to the nearest tenth if necessary.

1. 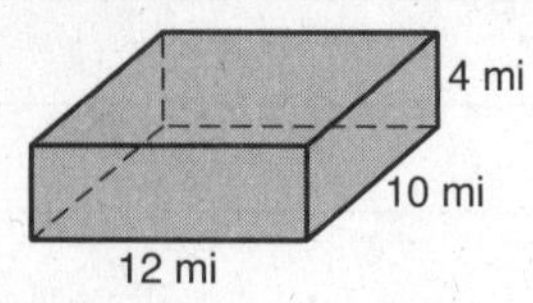

the rectangular prism

____

2. 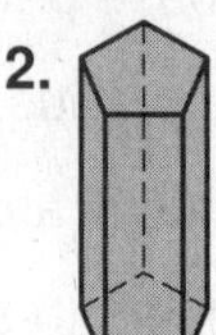

the regular pentagonal prism

____

3. a cube with edge length 20 inches ______________________

Find the lateral area and surface area of each right cylinder. Give your answers in terms of π.

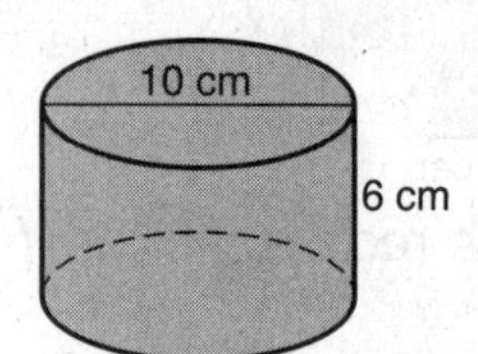

4. ______________________

5. a cylinder with base area 169π ft^2 and a height twice the radius ______________________

6. a cylinder with base circumference 8π m and a height one-fourth the radius ______________________

Find the surface area of each composite figure. Round to the nearest tenth.

7. 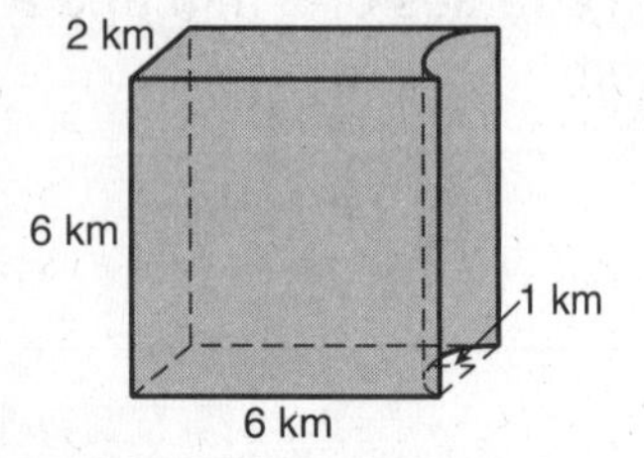

____

8. 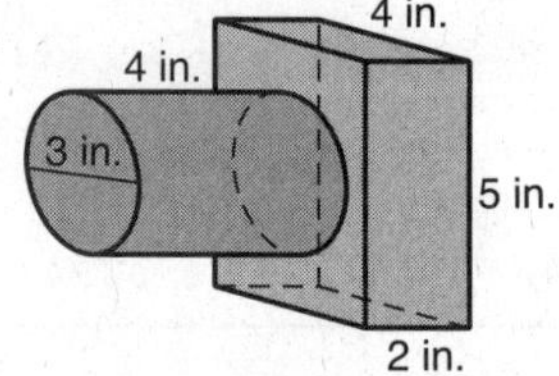

____

Describe the effect of each change on the surface area of the given figure.

9. 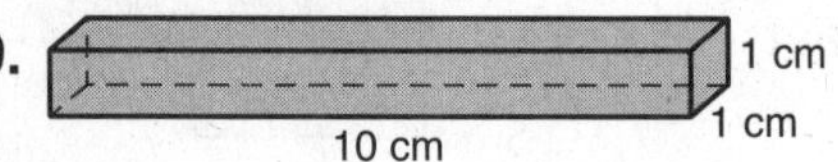

The dimensions are multiplied by 12.

____

10. 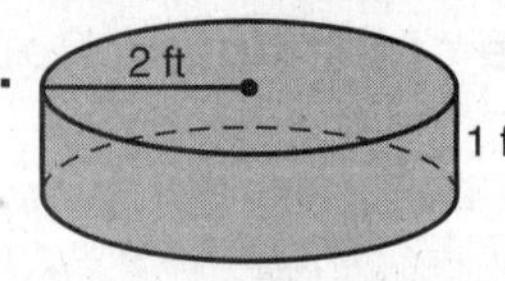

The dimensions are divided by 4.

____

Toby has eight cubes with edge length 1 inch. He can stack the cubes into three different rectangular prisms: 2-by-2-by-2, 8-by-1-by-1, and 2-by-4-by-1. Each prism has a volume of 8 cubic inches.

11. Tell which prism has the smallest surface-area-to-volume ratio. ______________________

12. Tell which prism has the greatest surface-area-to-volume ratio. ______________________

Copyright © by Holt, Rinehart and Winston.
All rights reserved.

Name ______________________ Date ___________ Class ___________

LESSON 10-4

Practice C

Surface Area of Prisms and Cylinders

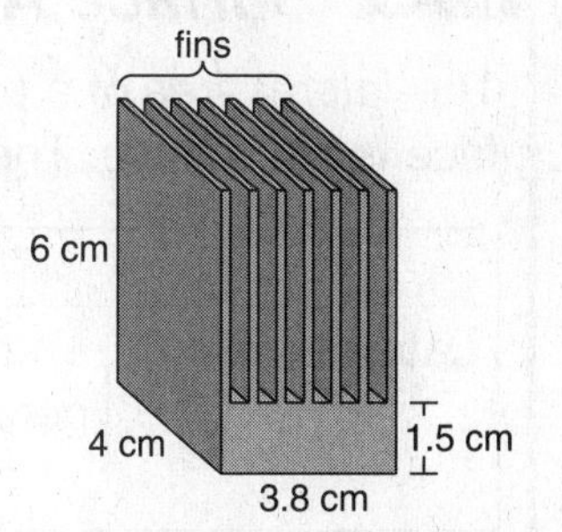

A heat sink is a chunk of metal that draws unwanted heat away from delicate electronic components and releases the heat into the air. The figure shows a typical heat sink for a desktop computer processor chip. Each fin is 2 mm wide and is 4 mm from the next fin.

1. Find the surface area of the heat sink in square millimeters.

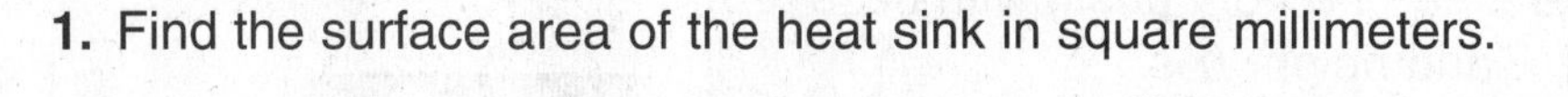

____

2. Explain why the heat sink has fins.

____

____

Find the surface area of each figure. Round to the nearest tenth if necessary.

3.

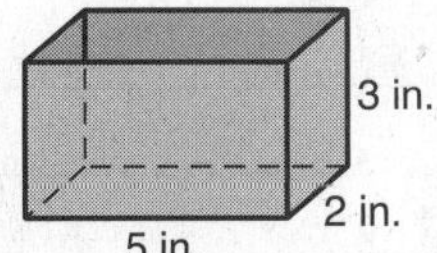

the rectangular prism with no top

____

4.

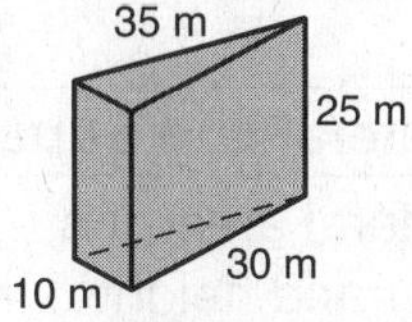

the right triangular prism

____

5.

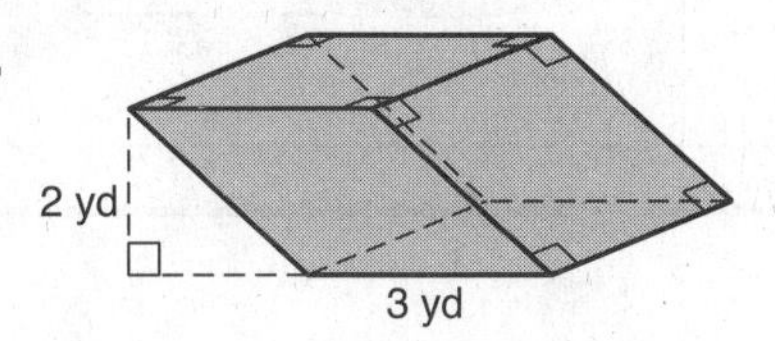

the oblique quadrilateral prism with each edge measuring 3 yards

____

6.

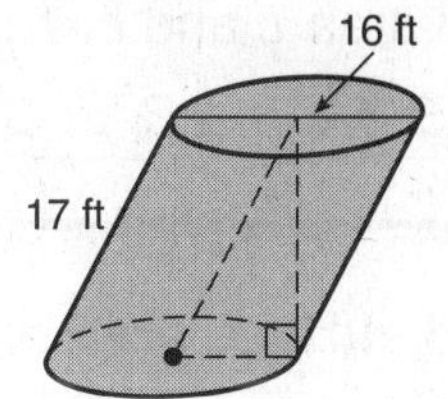

the oblique cylinder

____

7.

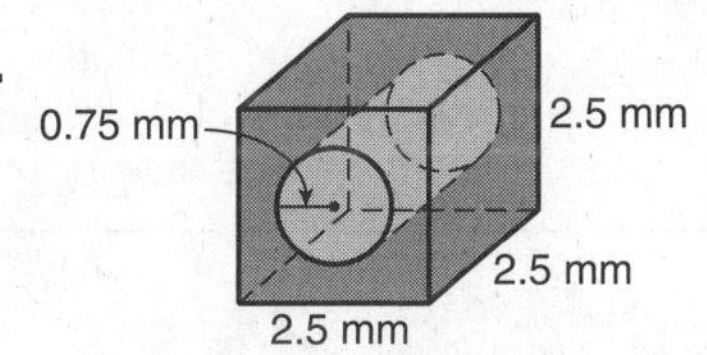

____

8.

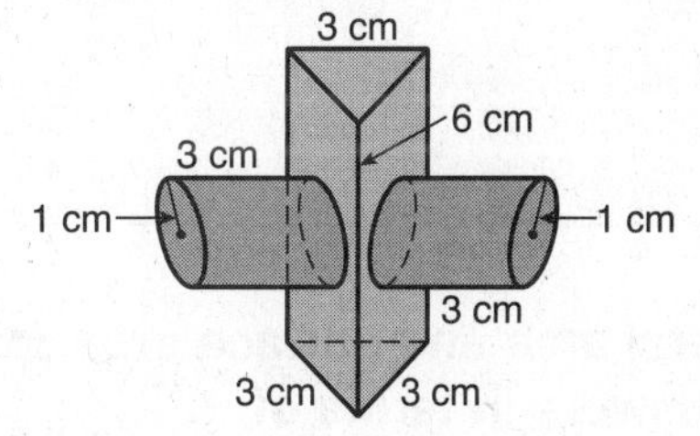

____

9. Draw a net of the oblique cylinder in Exercise 6.

Copyright © by Holt, Rinehart and Winston.
All rights reserved.

Name ______________________________ Date ____________ Class ____________

LESSON 10-4

Reteach

Surface Area of Prisms and Cylinders

The *lateral area* of a prism is the sum of the areas of all the *lateral faces*. A lateral face is not a base. The **surface area** is the total area of all faces.

Lateral and Surface Area of a Right Prism		
Lateral Area	The lateral area of a right prism with base perimeter P and height h is $L = Ph$.	h; lateral face
Surface Area	The surface area of a right prism with lateral area L and base area B is $S = L + 2B$, or $S = Ph + 2B$.	

The lateral area of a right cylinder is the curved surface that connects the two bases. The **surface area** is the total area of the curved surface and the bases.

Lateral and Surface Area of a Right Cylinder		
Lateral Area	The lateral area of a right cylinder with radius r and height h is $L = 2\pi rh$.	h; lateral surface
Surface Area	The surface area of a right cylinder with lateral area L and base area B is $S = L + 2B$, or $S = 2\pi rh + 2\pi r^2$.	

Find the lateral area and surface area of each right prism.

1.

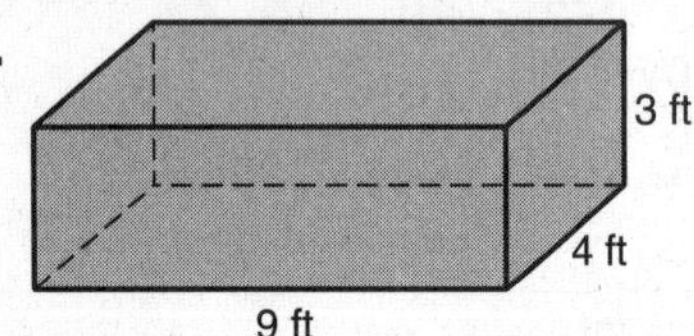

____

2.

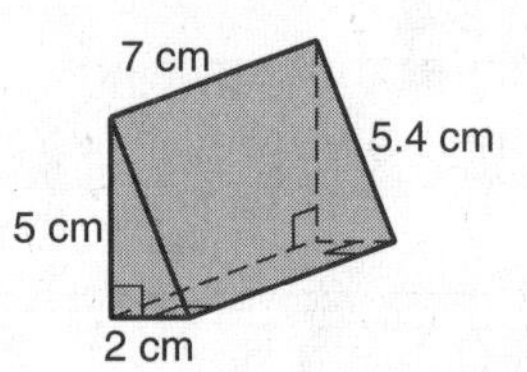

____

Find the lateral area and surface area of each right cylinder. Give your answers in terms of π.

3.

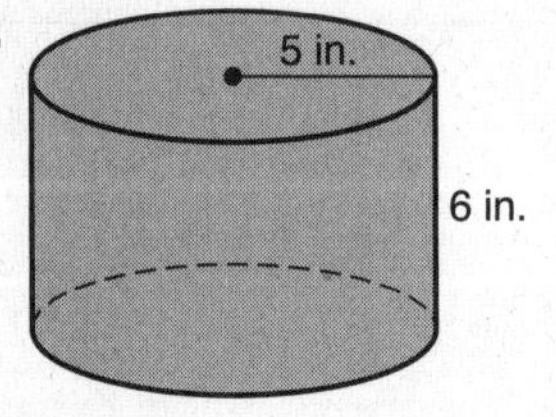

____

4.

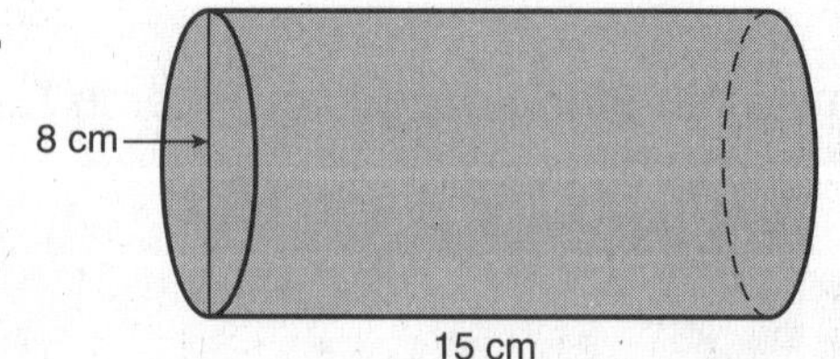

____

Copyright © by Holt, Rinehart and Winston.
All rights reserved.

Name ______________________________ Date __________ Class __________

LESSON 10-4

Reteach

Surface Area of Prisms and Cylinders continued

You can find the surface area of a composite three-dimensional figure like the one shown at right.

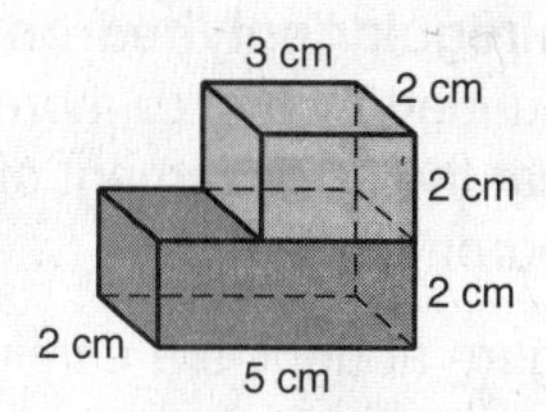

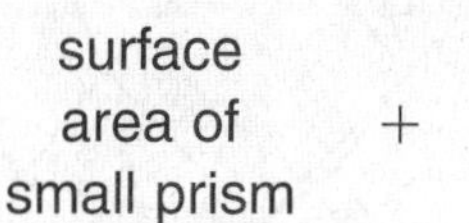

surface area of small prism + surface area of large prism − hidden surfaces

The dimensions are multiplied by 3. Describe the effect on the surface area.

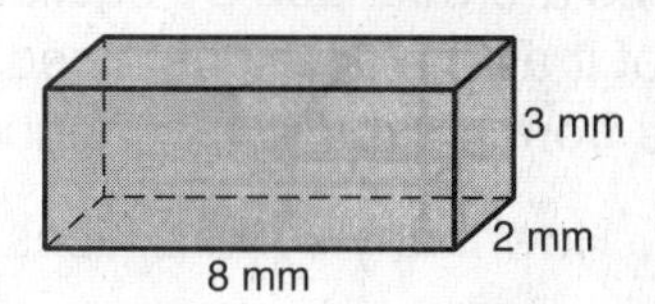

original surface area:new surface area, dimensions multiplied by 3:

original		new	
$S = Ph + 2B$		$S = Ph + 2B$	
$= 20(3) + 2(16)$	$P = 20$, $h = 3$, $B = 16$	$= 60(9) + 2(144)$	$P = 60$, $h = 9$, $B = 144$
$= 92 \text{ mm}^2$	Simplify.	$= 828 \text{ mm}^2$	Simplify.

Notice that $92 \cdot 9 = 828$. If the dimensions are multiplied by 3, the surface area is multiplied by 3^2, or 9.

Find the surface area of each composite figure. Be sure to subtract the hidden surfaces of each part of the composite solid. Round to the nearest tenth.

5.

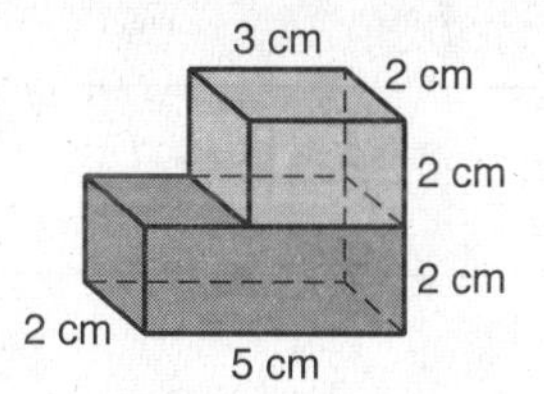

____

6.

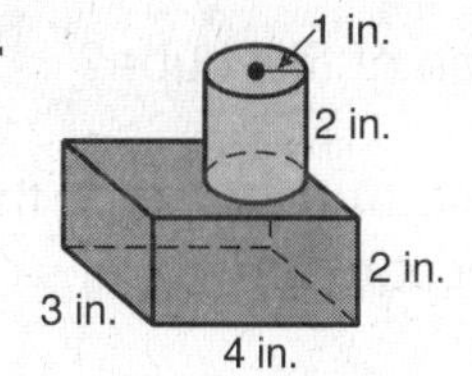

____

Describe the effect of each change on the surface area of the given figure.

7. The length, width, and height are multiplied by 2.

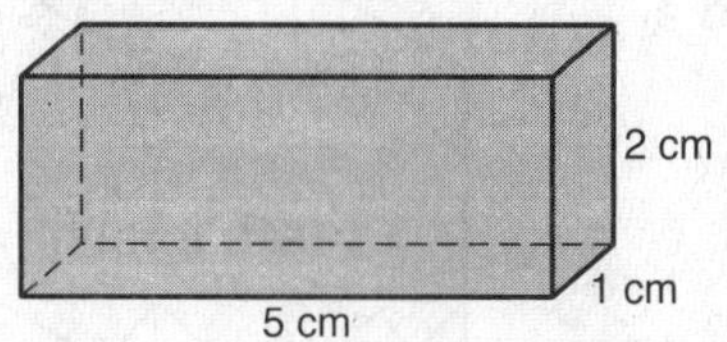

____

8. The height and radius are multiplied by $\frac{1}{2}$.

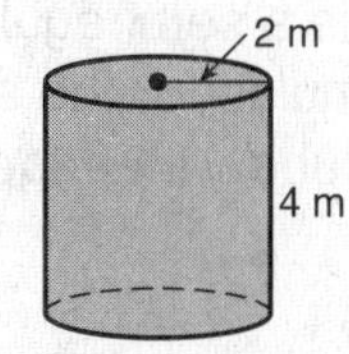

____

Copyright © by Holt, Rinehart and Winston.
All rights reserved.

Name ______________________ Date ____________ Class ____________

LESSON 10-4

Challenge

Surface Area and Volume of Semiregular Polyhedra

A **semiregular polyhedron** is a convex polyhedron whose faces are bounded by two or more types of regular polygons in such a way that the arrangement of polygons at each vertex of the polyhedron is identical.

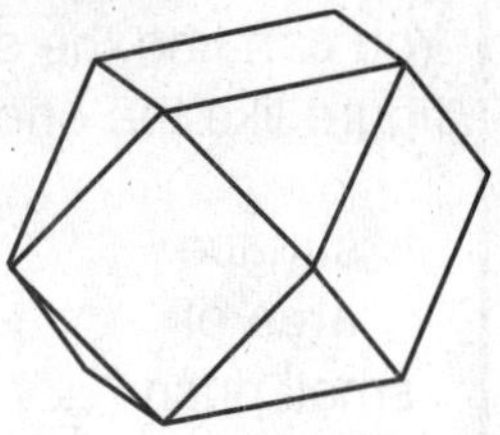

The figure at right is a semiregular polyhedron called a *cuboctahedron.* Its faces are bounded by equilateral triangles and squares. You can think of it as the figure obtained if you "cut off" eight congruent pieces from a cube in the manner shown.

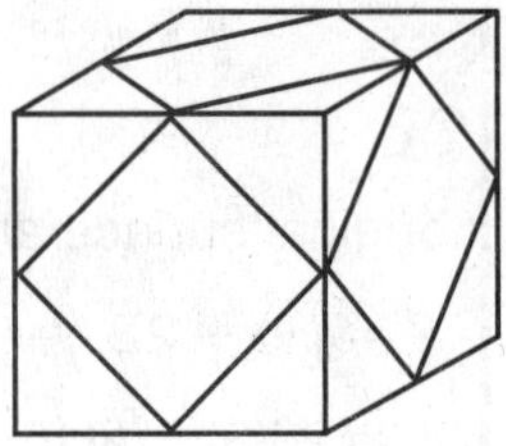

1. a. How many square faces does a cuboctahedron have? ____________

   b. How many triangular faces does a cuboctahedron have? ____________

2. What type of figure is each piece that is "cut off" from the original cube? ____________

Suppose that the length of each edge of a cuboctahedron is 10 inches.

3. a. What is the area of each square face? ____________

   b. What is the area of each triangular face? ____________

   c. What is the total surface area? ____________

4. a. What is the length of each edge of the original cube from which the cuboctahedron was "cut"? ____________

   b. What is the volume of this cube? ____________

   c. What is the volume of each piece that was "cut off" from the cube? ____________

   d. What is the volume of the cuboctahedron? ____________

5. Generalize your results from Exercises 3 and 4 to write formulas for the surface area S and volume V of a cuboctahedron with edge of length n.

   $S =$ ____________ $V =$ ____________

6. When eight congruent pieces are cut from a cube in the manner shown at right, the result is a semiregular polyhedron called a *truncated cube.* Write formulas for the surface area S and volume V of a truncated cube with edge of length m.

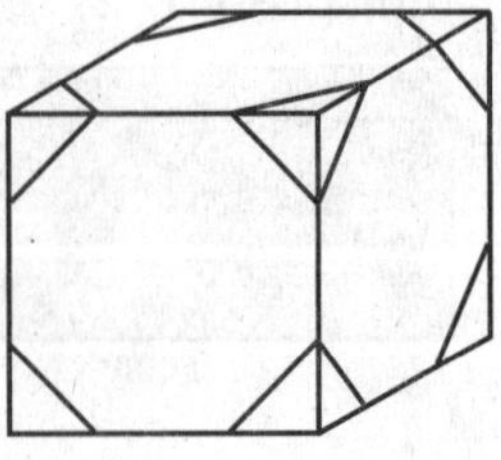

   $S =$ ____________

   $V =$ ____________

Copyright © by Holt, Rinehart and Winston.
All rights reserved.

Name ______________________ Date __________ Class __________

LESSON 10-4

Problem Solving

Surface Area of Prisms and Cylinders

1. The lateral area of the regular pentagonal prism below is 220 mm^2. What is the surface area? Round to the nearest tenth if necessary.

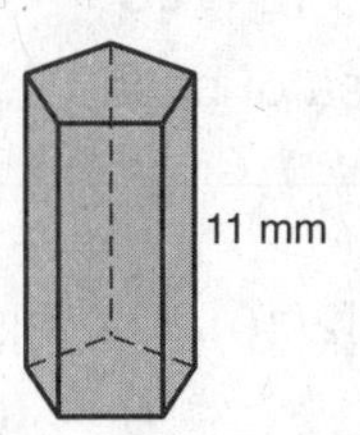

____

2. A sheet of metal 8 feet long and 6 feet wide is to be cut into cylindrical cans like the one shown. How many lateral surfaces for the cans can be cut from the metal with as little waste as possible?

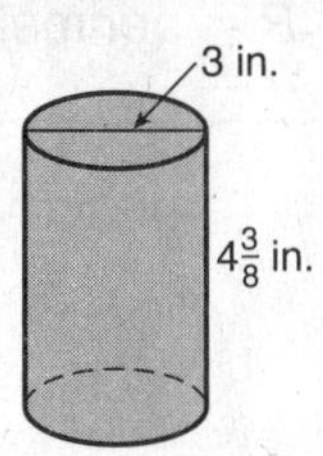

____

Choose the best answer.

3. The surface area of a cube is increased so that it is 9 times its original surface area. How did the length of the cube change?

   A The length was doubled.
   B The length was tripled.
   C The length was quadrupled.
   D The length was multiplied by 9.

4. A rectangular prism has a surface area of 152 square inches. If the length, width, and height are all changed to $\frac{1}{2}$ their original size, what will be the new surface area of the prism?

   F 19 in^2
   G 38 in^2
   H 76 in^2
   J 114 in^2

5. Determine the surface area exposed to the air of the composite figure shown. Round to the nearest tenth.

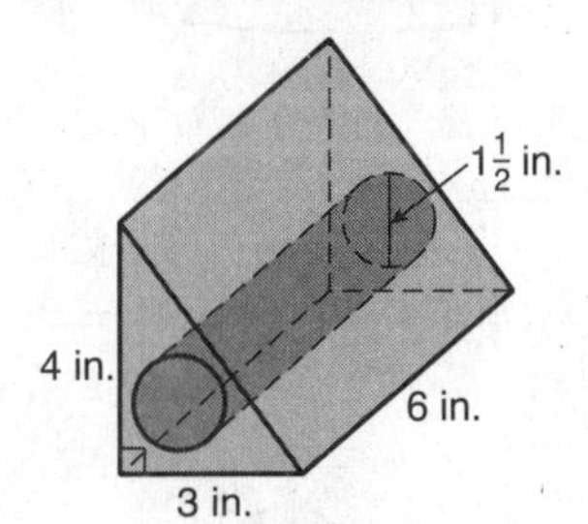

   A 98.1 in^2
   B 107.6 in^2
   C 108.7 in^2
   D 110.5 in^2

6. Which of the two cylindrical cans has a greater surface area?

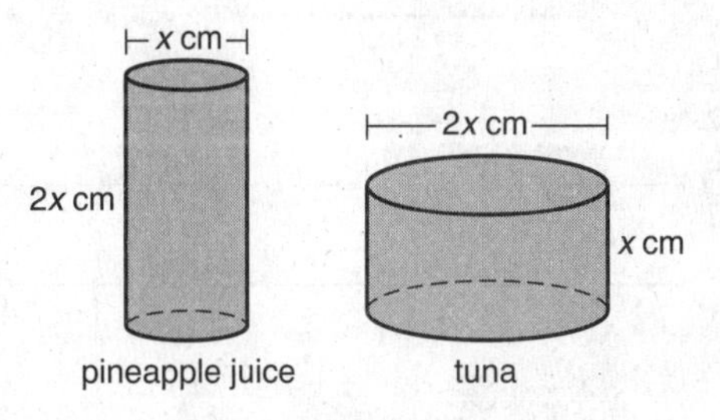

   F pineapple juice can
   G tuna can
   H The two cans have the same surface area.
   J It is impossible to determine which can has a greater surface area.

Copyright © by Holt, Rinehart and Winston.
All rights reserved.

Name ______________________ Date ____________ Class ____________

LESSON 10-4

Reading Strategies

Use a Concept Map

Use the concept maps below to help you understand and use lateral and surface area formulas.

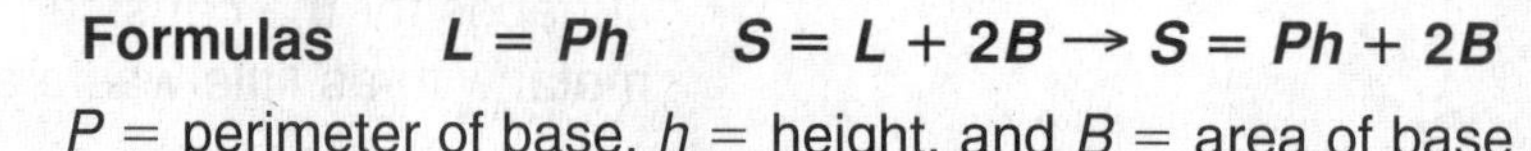

Formulas $L = Ph$ $\quad S = L + 2B \rightarrow S = Ph + 2B$

P = perimeter of base, h = height, and B = area of base

Lateral Area L and Surface Area S of Right Prisms

Diagram

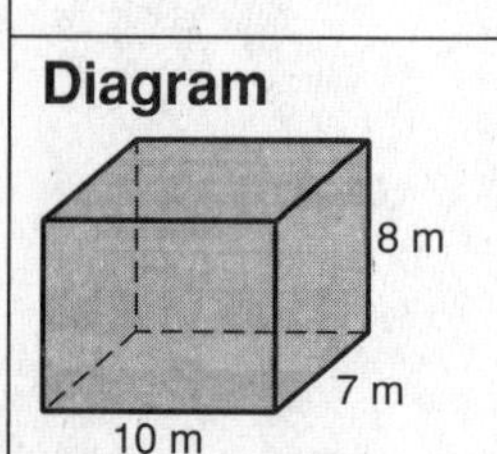

Examples

$L = [2(10) + 2(7)](8) = \mathbf{272}\text{ cm}^2$

$B = 10(7) = 70\text{ cm}^2$

$2B = \mathbf{140}\text{ cm}^2$

$\mathbf{S = 272 + 140} = 412\text{ cm}^2$

Formulas $L = 2\pi rh$ $\quad S = L + 2B \rightarrow S = 2\pi rh + 2\pi r^2$

r = radius, h = height, and B = area of base

Lateral Area L and Surface Area S of Right Cylinders

Diagram

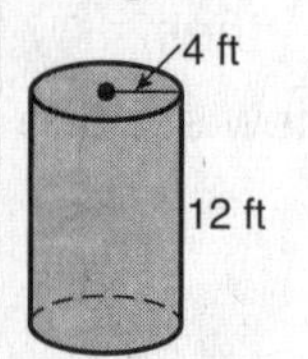

Examples

$L = 2\pi(4)(12) = \mathbf{96\pi}\text{ ft}^2$

$B = \pi(4)^2 = 16\pi\text{ ft}^2$

$2B = \mathbf{32\pi}\text{ ft}^2$

$\mathbf{S = 96\pi + 32\pi} \approx 402.1\text{ ft}^2$

Find the lateral area and surface area of each figure. Round to the nearest tenth if necessary.

1.

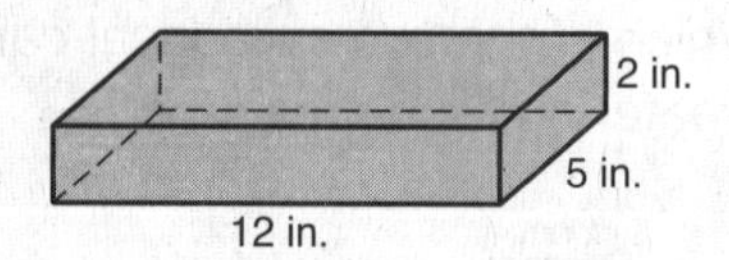

L = ______________________

S = ______________________

2.

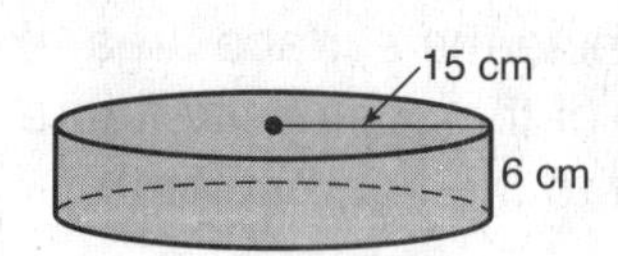

L = ______________________

S = ______________________

3.

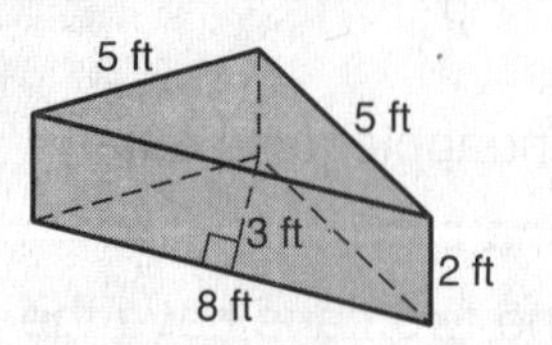

L = ______________________

S = ______________________

4.

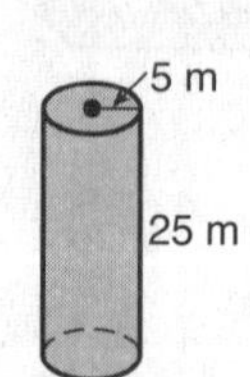

L = ______________________

S = ______________________

Copyright © by Holt, Rinehart and Winston.
All rights reserved.

Name ________________ Date __________ Class __________

LESSON 10-5

Practice A

Surface Area of Pyramids and Cones

Write each formula.

1. lateral area of a regular pyramid with base perimeter P and slant height ℓ ________________

2. lateral area of a right cone with radius r and slant height ℓ ________________

3. surface area of a regular pyramid with lateral area L and base area B ________________

4. surface area of a right cone with lateral area L and base area B ________________

Find the lateral area and surface area of each regular pyramid. Round to the nearest tenth if necessary.

5. 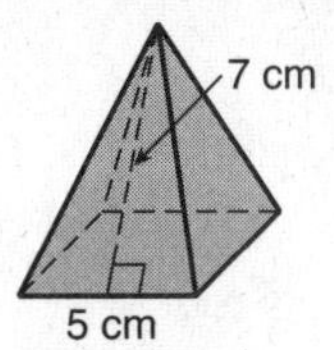

the regular square pyramid

____

6. 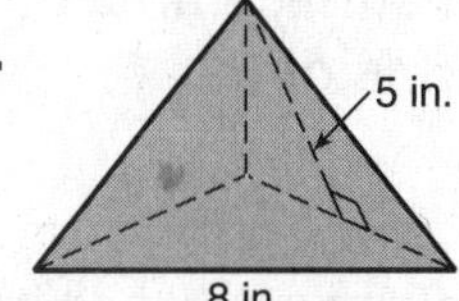

the regular triangular pyramid

____

Find the lateral area and surface area of each right cone. Give your answers in terms of π.

7. 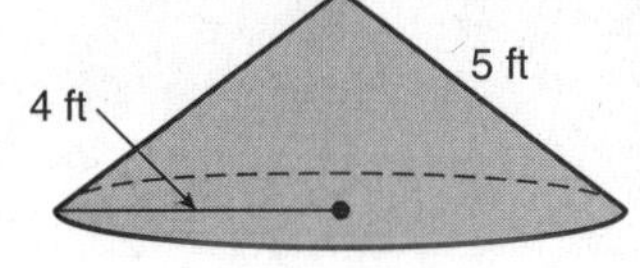

____

8. a right cone with radius 3 m and slant height 12 m

____

Complete Exercises 9–11 to describe the effect on the surface area of dividing the dimensions of a cone by 2. Give your answers in terms of π.

9. Find the surface area of a right cone with radius 2 yards and slant height 6 yards. ________________

10. Find the surface area of a right cone with radius 1 yard and slant height 3 yards. ________________

11. Describe the effect on the surface area of dividing the dimensions of a right cone by 2.

____

12. Find the surface area of the composite figure in terms of π.

____

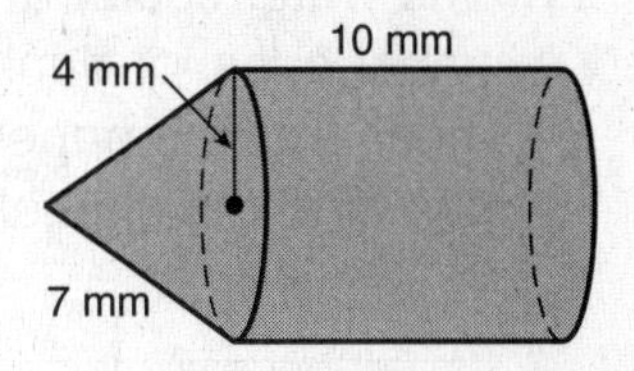

Copyright © by Holt, Rinehart and Winston.
All rights reserved.

Name ______________________ Date __________ Class __________

LESSON 10-5

Practice B

Surface Area of Pyramids and Cones

Find the lateral area and surface area of each regular right solid. Round to the nearest tenth if necessary.

1.

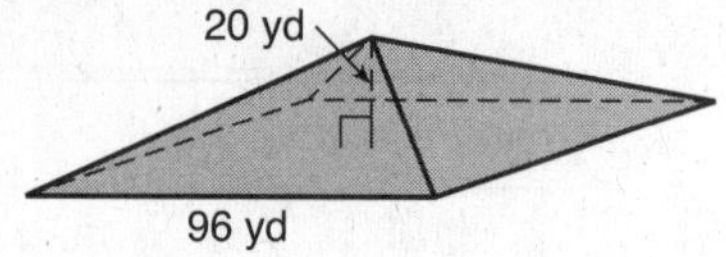

2.

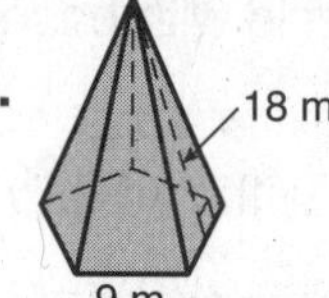

_________________________ _________________________

3. a regular hexagonal pyramid with base edge length 12 mi and slant height 15 mi

____

Find the lateral area and surface area of each right cone. Give your answers in terms of π.

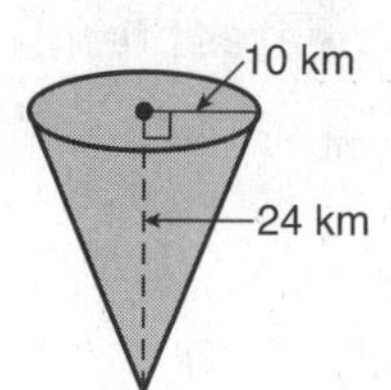

4. _________________________

5. a right cone with base circumference 14π ft and slant height 3 times the radius _________________________

6. a right cone with diameter 240 cm and altitude 35 cm

____

Describe the effect of each change on the surface area of the given figure.

7.

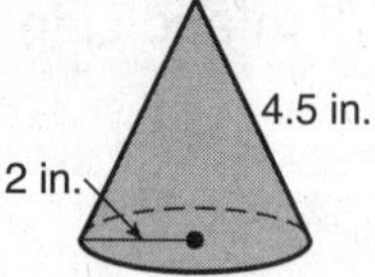

The dimensions are multiplied by $\frac{1}{5}$.

____

8.

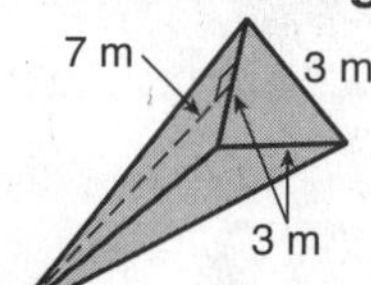

The dimensions are multiplied by $\frac{3}{2}$.

____

Find the surface area of each composite figure. Round to the nearest tenth if necessary.

9.

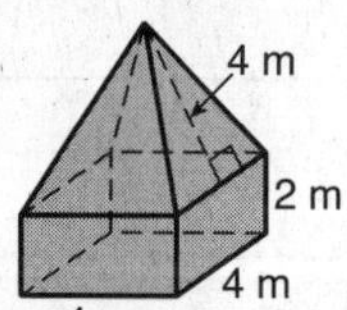

____

10.

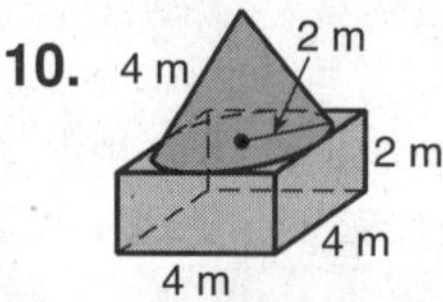

____

11. The water cooler at Mohammed's office has small conical paper cups for drinking. He uncurls one of the cups and measures the paper. Based on the diagram of the uncurled cup, find the diameter of the cone.

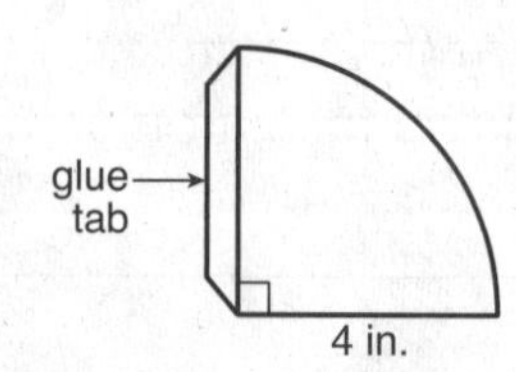

____

Copyright © by Holt, Rinehart and Winston.
All rights reserved.

Name ______________________ Date __________ Class __________

LESSON 10-5

Practice C

Surface Area of Pyramids and Cones

Use the figure for Exercises 1–3. The figure shown can be curled into an open-topped cone.

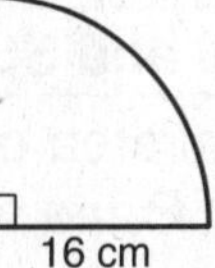

1. Find the radius of the top of the cone. __________

2. Find the radius of a circle that has the same area as the lateral area of the cone. __________

3. Name the mathematical relationship between the answer to Exercise 2 and the radius and slant height of the cone. __________

Find the radius of a circle that has the same area as the lateral area of each cone described in Exercises 4–6. Give exact answers.

4. $\ell = 32$ ft; $r = 8$ ft

5. $\ell = 68$ m; $r = 17$ m

6. $\ell = 5$ cm; $r = 2$ cm

__________ __________ __________

7. A cone with an open base can be formed from any partial circle. Develop a formula for the lateral area of a cone based on ℓ, the slant height, and d, the degree measure of the interior angle in the partial circle. (For instance, d in the figure above equals 90.) __________

The figures below can be curled into open frustums. Find the lateral area of the outside and the radius of both bases in each frustum. Round to the nearest tenth.

8.

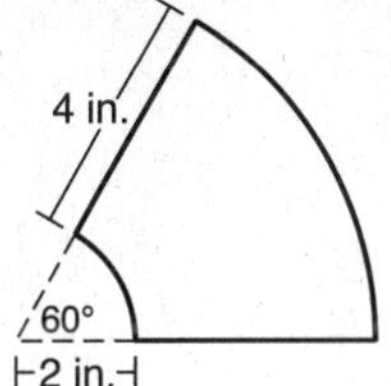

9.

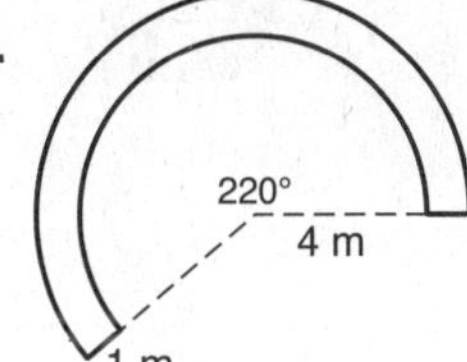

__________ __________

10. Terry is driving through the desert when she notices the engine is low on oil. She has a few quarts of motor oil in the trunk of the car, but she does not have a funnel. Fortunately, Terry finds a piece of $8\frac{1}{2}$-by-11-inch notebook paper in the backseat. She wants to cut the paper to make a funnel with a 1-inch diameter hole on the bottom and the longest slant height possible. Find the diameter of the top of the funnel. Draw the pattern Terry will cut out before curling up the funnel. __________

Copyright © by Holt, Rinehart and Winston.
All rights reserved.

Name ____________________ Date __________ Class __________

LESSON 10-5

Reteach

Surface Area of Pyramids and Cones

Lateral and Surface Area of a Regular Pyramid		
Lateral Area	The lateral area of a regular pyramid with perimeter P and slant height ℓ is $L = \frac{1}{2}P\ell$.	slant height; ℓ; base
Surface Area	The surface area of a regular pyramid with lateral area L and base area B is $S = L + B$, or $S = \frac{1}{2}P\ell + B$.	

Lateral and Surface Area of a Right Cone		
Lateral Area	The lateral area of a right cone with radius r and slant height ℓ is $L = \pi r\ell$.	slant height; ℓ; r; base
Surface Area	The surface area of a right cone with lateral area L and base area B is $S = L + B$, or $S = \pi r\ell + \pi r^2$.	

Find the lateral area and surface area of each regular pyramid. Round to the nearest tenth.

1.

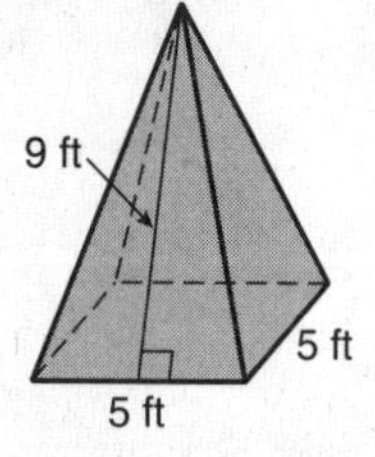

2.

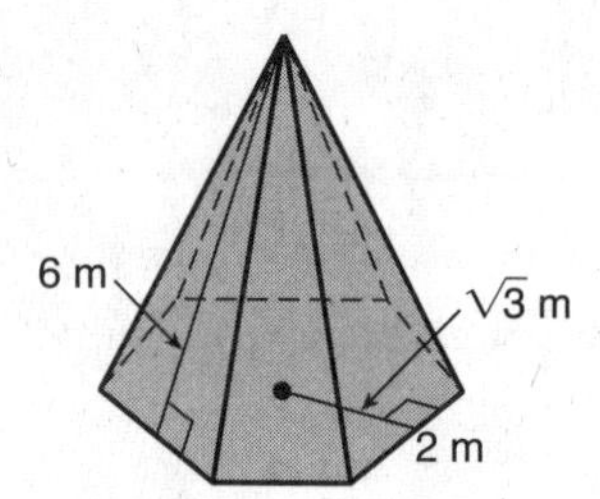

________________________ ________________________

Find the lateral area and surface area of each right cone. Give your answers in terms of π.

3.

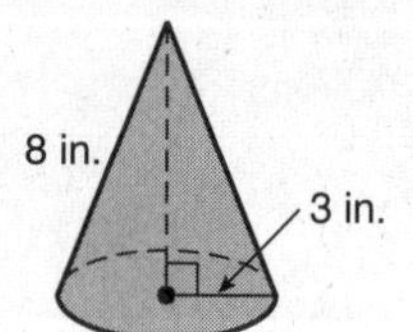

4.

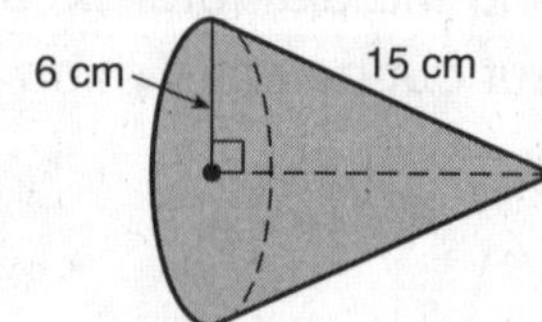

________________________ ________________________

Copyright © by Holt, Rinehart and Winston.
All rights reserved.

Name ______________________ Date __________ Class __________

LESSON 10-5

Reteach

Surface Area of Pyramids and Cones continued

The radius and slant height of the cone at right are doubled. Describe the effect on the surface area.

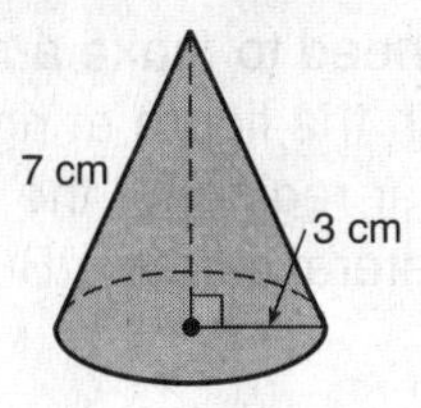

original surface area:

$S = \pi r\ell + \pi r^2$

$= \pi(3)(7) + \pi(3)^2$ $\quad r = 3, \ell = 7$

$= 30\pi \text{ cm}^2$ $\quad$ Simplify.

new surface area, dimensions doubled:

$S = \pi r\ell + \pi r^2$

$= \pi(6)(14) + \pi(6)^2$ $\quad r = 6, \ell = 14$

$= 120\pi \text{ cm}^2$ $\quad$ Simplify.

If the dimensions are doubled, then the surface area is multiplied by 2^2, or 4.

Describe the effect of each change on the surface area of the given figure.

5. The dimensions are tripled.

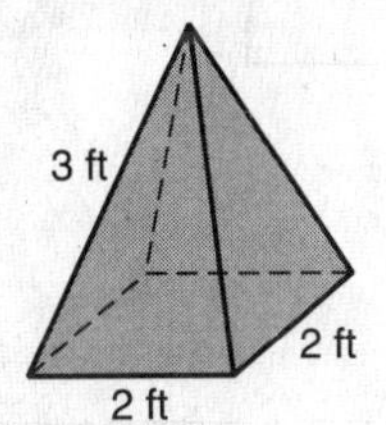

____

6. The dimensions are multiplied by $\frac{1}{2}$.

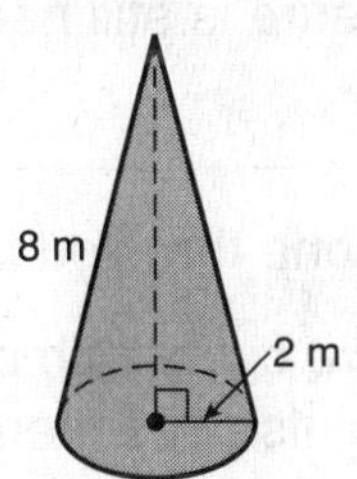

____

Find the surface area of each composite figure.

7. *Hint:* Do not include the base area of the pyramid or the upper surface area of the rectangular prism.

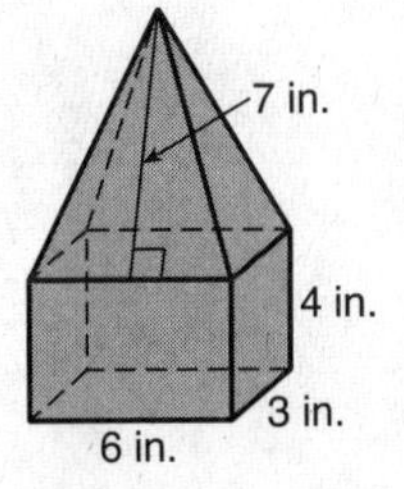

____

8. *Hint:* Add the lateral areas of the cones.

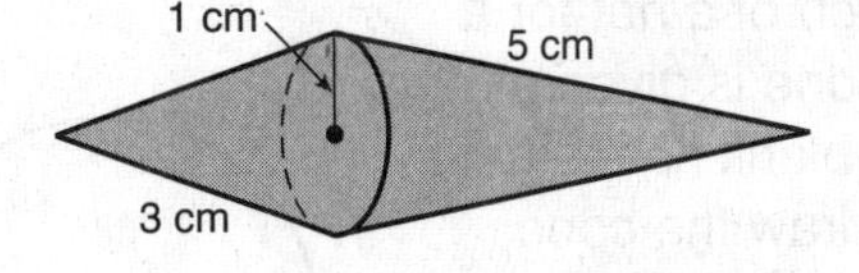

____

Copyright © by Holt, Rinehart and Winston.
All rights reserved.

Name ________________________ Date ____________ Class ____________

LESSON 10-5

Challenge

Making Nets for Right Cones

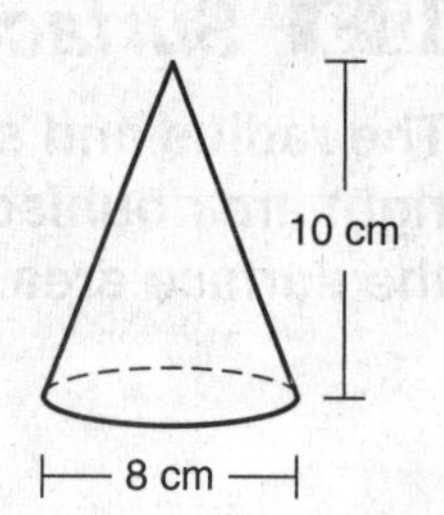

Suppose that you need to make a model of a cone that has the dimensions given in the figure at right. You know that the net for the cone consists of a circular region for the base and a region bounded by a sector of a circle for the lateral surface. But how do you know the exact size of each piece?

Give each measure for the right cone shown at right. When necessary, round to the nearest tenth of a centimeter.

1. radius ________________ 2. circumference ________________

3. height ________________ 4. slant height ________________

5. A sketch of a net for the cone shown above is given at right.

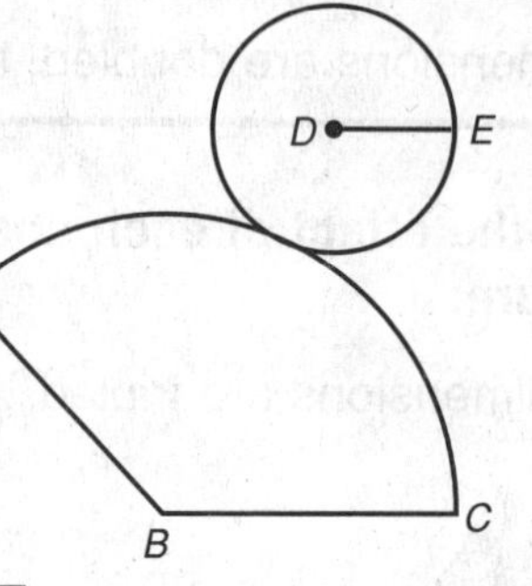

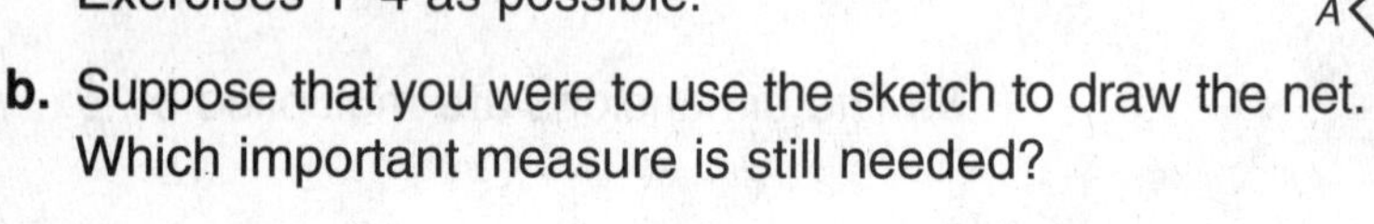

   a. Label the sketch with as many of the measures from Exercises 1–4 as possible.

   b. Suppose that you were to use the sketch to draw the net. Which important measure is still needed?

   ____

6. Refer to the net for the cone that you labeled in Exercise 5.

   a. Suppose that the *entire* circle with center at point *B* was drawn. What would be its circumference? ________________

   b. What is the length of the arc that is drawn from *A* to *C*? ________________

   c. What percent of the entire circle is the arc from *A* to *C*? ________________

   d. Multiply 360° by your percent from part **c**. What is the measure of $\angle ABC$, rounded to the nearest whole degree? ________________

7. Refer to your results from Exercises 5 and 6. Using a compass, ruler, and protractor, draw an accurate real-size net for the cone. Then assemble the net to make a model of the cone.

8. A sketch of a net for a right cone is given at right. In the blank space to its right, draw the cone, making the height and diameter of the cone in the drawing proportional to the actual height and diameter. Be sure to label the height and diameter.

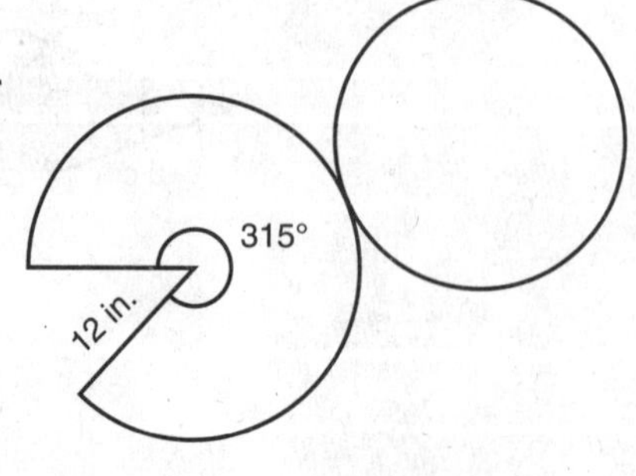

Copyright © by Holt, Rinehart and Winston.
All rights reserved.

Name ______________________ Date __________ Class __________

LESSON **10-5**

Problem Solving

Surface Area of Pyramids and Cones

1. Find the diameter of a right cone with slant height 18 centimeters and surface area 208π square centimeters.

____

2. Find the surface area of a regular pentagonal pyramid with base area 49 square meters and slant height 13 meters. Round to the nearest tenth.

____

3. A piece of paper in the shape shown is folded to form a cone. What is the diameter of the base of the cone that is formed? Round to the nearest tenth.

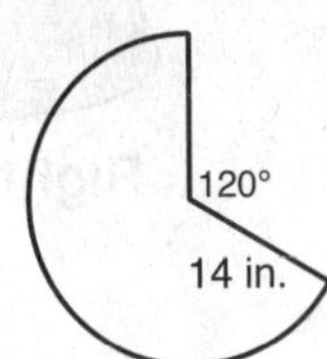

____

4. The right cone has a surface area of 240π square millimeters. What is the radius of the cone?

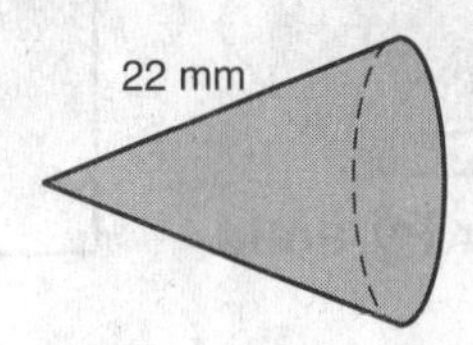

____

Choose the best answer.

5. A square pyramid has a base with a side length of 9 centimeters and a slant height that is 4 centimeters more than $1\frac{1}{2}$ times the length of the base. Find the surface area of the pyramid.

   A 162 cm^2 **C** 315 cm^2
   B 243 cm^2 **D** 396 cm^2

6. A cone has a surface area of 64π square inches. If the radius and height are each multiplied by $\frac{3}{4}$, what will be the new surface area of the cone?

   F 36π in^2 **H** 60π in^2
   G 48π in^2 **J** 96π in^2

7. Find the surface area of the composite figure. Round to the nearest tenth.

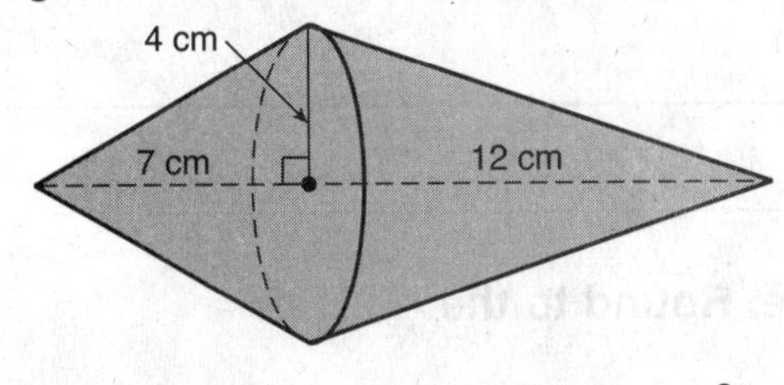

   A 238.8 cm^2 **C** 311.0 cm^2
   B 260.3 cm^2 **D** 361.3 cm^2

8. A cone has a base diameter of 6 yards. What is the slant height of the cone if it has the same surface area as the square pyramid shown? Round to the nearest tenth.

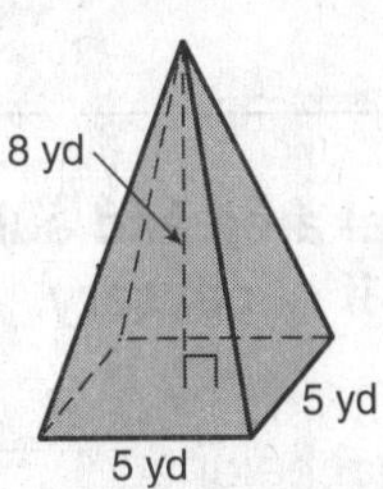

   F 8.1 yd **H** 11.3 yd
   G 8.5 yd **J** 25.6 yd

Copyright © by Holt, Rinehart and Winston.
All rights reserved.

Name ______________________ Date __________ Class __________

LESSON 10-5

Reading Strategies

Compare and Contrast

The diagram below summarizes the similarities and differences between regular pyramids and right cones and their lateral and surface areas.

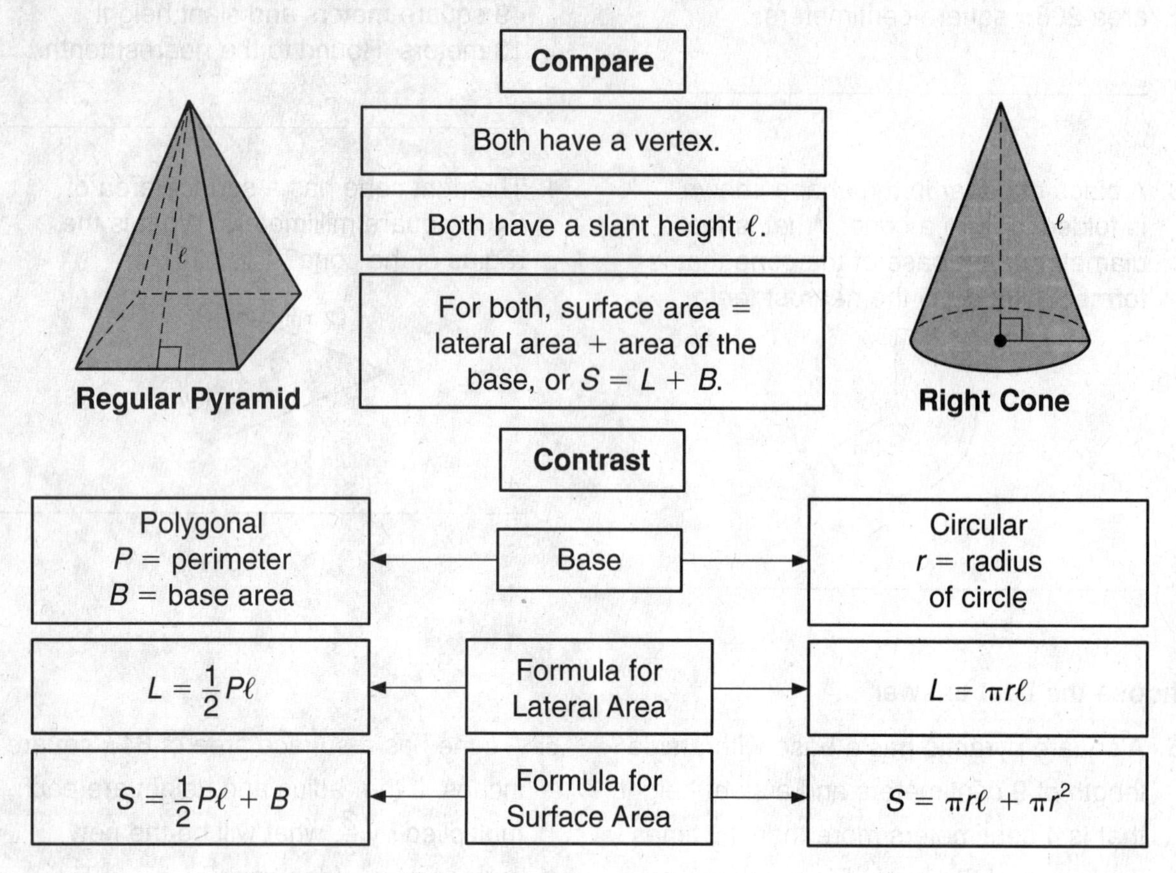

Answer the following.

1. Look at the pyramid and cone above. Why do you think the slant height is so named?

____

2. Look at the formulas for surface area for each figure. Why do you think the formula for the pyramid uses B for area of the base and the formula for the cone does not?

____

____

Find the lateral area and surface area of each figure. Round to the nearest tenth if necessary.

3. 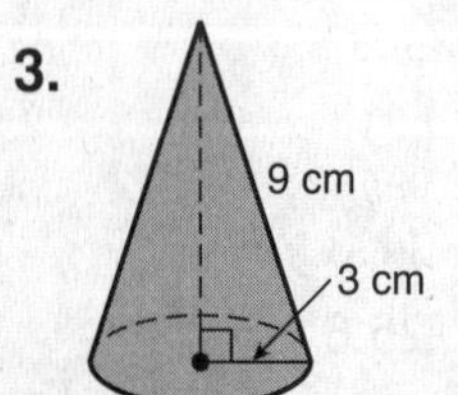

L = ______________

S = ______________

4. 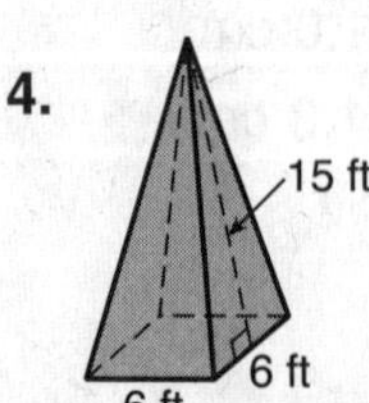

L = ______________

S = ______________

Copyright © by Holt, Rinehart and Winston.
All rights reserved.

Name ______________________ Date __________ Class __________

LESSON **10-6**

Practice A

Volume of Prisms and Cylinders

Write each formula.

1. volume of a cube with edge length *s* ____________
2. volume of a prism with base area *B* and height *h* ____________
3. volume of a cylinder with radius *r* and height *h* ____________
4. volume of a right rectangular prism with length ℓ, width *w*, and height *h* ____________

Find the volume of each prism. Round to the nearest tenth if necessary.

5. 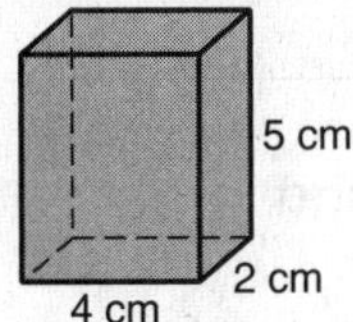

the right rectangular prism

____

6. 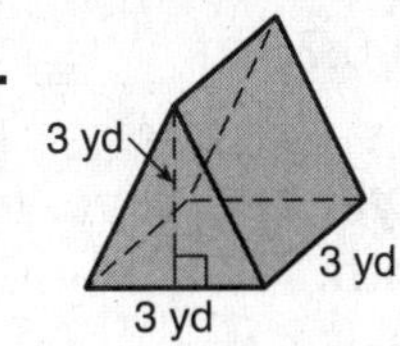

the triangular prism

____

7. Laetitia needs to store 8 boxes while she is moving. Each box is a cube with edge length 3 feet. A storage facility charges $0.75 for every cubic foot of storage per month. Find the amount of money Laetitia will pay to store her boxes for one month. ____________

Find the volume of each cylinder. Give your answers both in terms of π and rounded to the nearest tenth.

8. 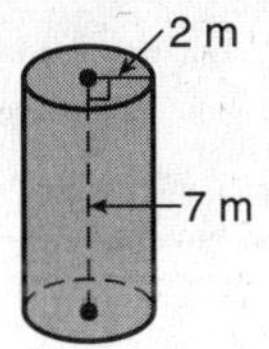

____

9. a cylinder with diameter 20 in. and height 2 in.

____

Complete Exercises 10–12 to describe the effect on the volume of multiplying each dimension of a prism by 3.

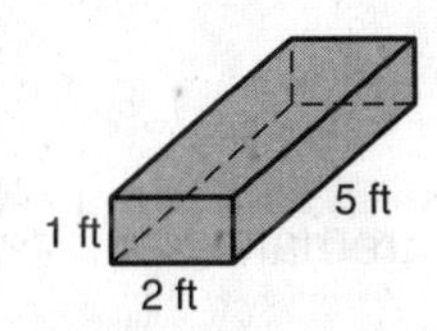

10. Find the volume of the prism. ____________
11. Find the volume of the prism after each dimension is multiplied by 3. ____________
12. Describe the effect on the volume of multiplying each dimension of a prism by 3.

____

13. Find the volume of the composite figure. Round to the nearest tenth.

____

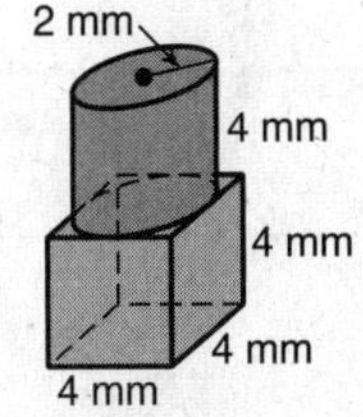

Copyright © by Holt, Rinehart and Winston.
All rights reserved.

Name ______________________ Date ____________ Class ____________

LESSON 10-6

Practice B

Volume of Prisms and Cylinders

Find the volume of each prism. Round to the nearest tenth if necessary.

1. 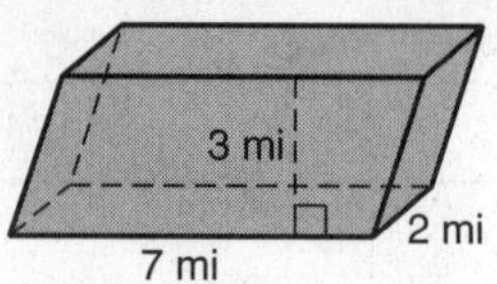

the oblique rectangular prism

____

2. 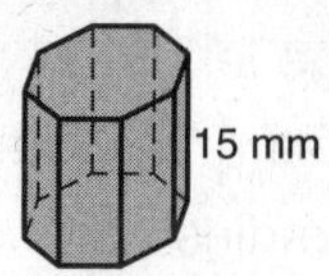

the regular octagonal prism

____

3. a cube with edge length 0.75 m ________________________

Find the volume of each cylinder. Give your answers both in terms of π and rounded to the nearest tenth.

4. 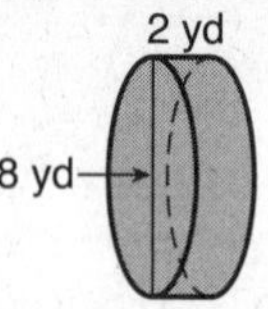

____

5. 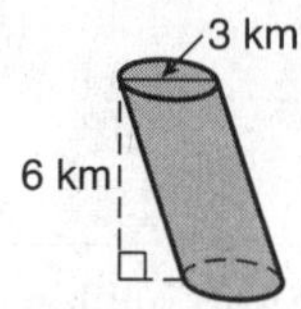

____

6. a cylinder with base circumference 18π ft and height 10 ft ________________________

7. CDs have the dimensions shown in the figure. Each CD is 1 mm thick. Find the volume in cubic centimeters of a stack of 25 CDs. Round to the nearest tenth.

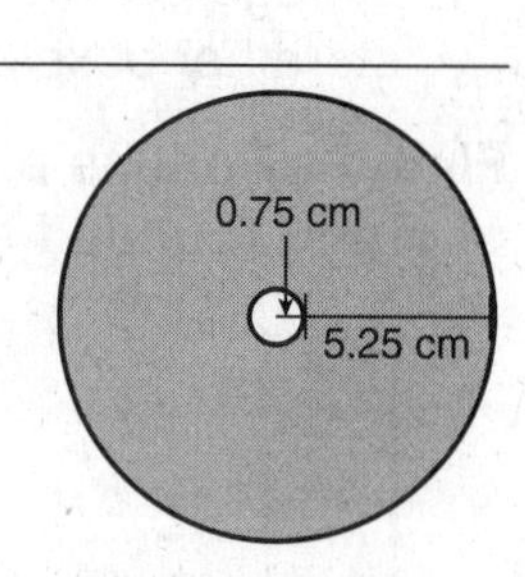

____

Describe the effect of each change on the volume of the given figure.

8. 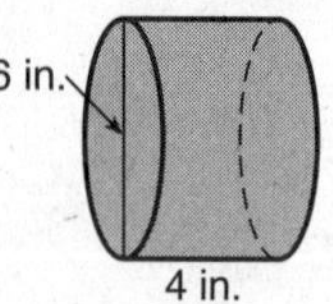

The dimensions are halved.

____

9. 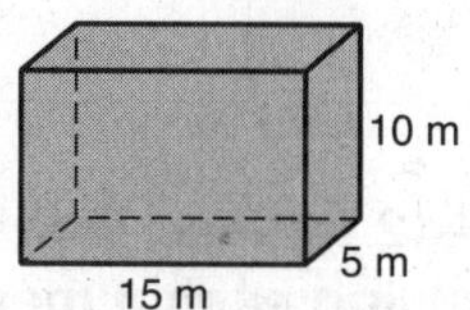

The dimensions are divided by 5.

____

Find the volume of each composite figure. Round to the nearest tenth.

10. 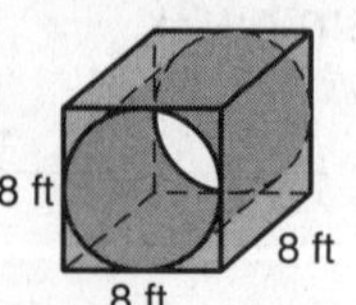

____

11. 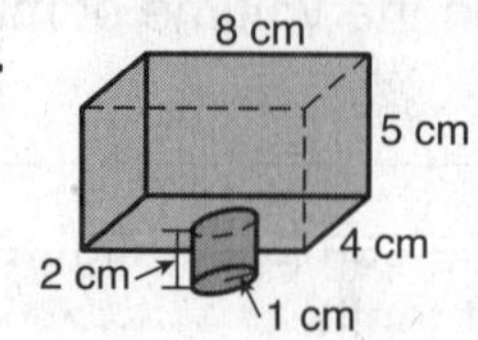

____

Copyright © by Holt, Rinehart and Winston.
All rights reserved.

Name ______________________ Date __________ Class __________

LESSON 10-6

Practice C

Volume of Prisms and Cylinders

1. Find the volume-to-surface-area ratio for these two cylinders. Round to the nearest tenth.

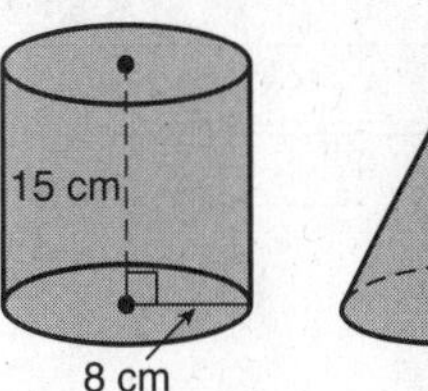

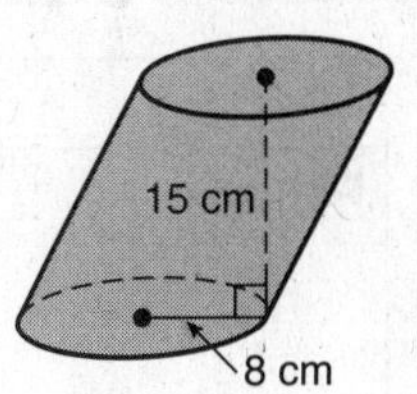

____

A chocolate bar is in the shape of a rectangular prism with length 5 in., width $2\frac{1}{4}$ in., and height $\frac{1}{4}$ in. The bar weighs 1.75 ounces. The chart shows some of the nutritional information for the chocolate bar. Round your answers to Exercises 2–4 to the nearest hundredth.

Serving size $\frac{1}{2}$ bar
Calories 135 **Total fat** 8 g (12% DV) **Total carb** 14 g (5% DV)

2. Find the density of the chocolate bar (ounces/cubic inch). ________________________

3. Find the volume of chocolate that contains 100 calories. ________________________

4. The "% DV" indicates the percentage of the recommended daily amount for that nutrient. Find the volume of chocolate that would provide 100% of the recommended daily amount of carbohydrates. (*Note:* This is NOT a healthy diet.) ________________________

In the sciences, quantities of liquids are measured in liters and milliliters. One milliliter of water has the same volume as a cube with edge length 1 cm.

5. Tell what size cube has the same volume as 1 liter of water.

____

6. In a science lab, liquids are often measured out in tall, thin cylinders called graduated cylinders. One graduated cylinder has a diameter of 2 centimeters, and 8 milliliters of water are poured into it. Tell how high the water will reach. Round to the nearest tenth. ________________________

Find the volume of each figure. Round to the nearest tenth.

7. 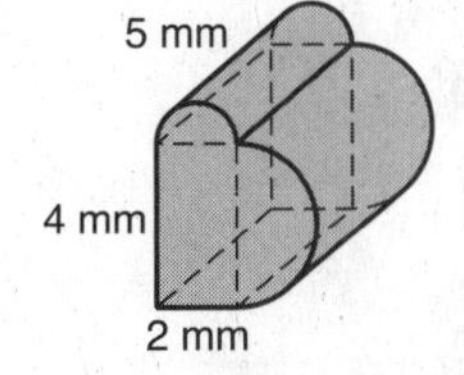

8. 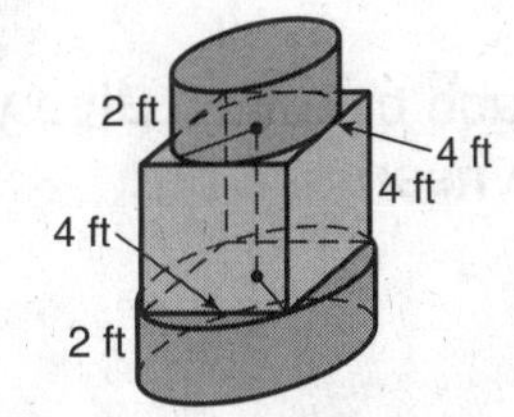

________________________ ________________________

Copyright © by Holt, Rinehart and Winston.
All rights reserved.

Name ________________________ Date ____________ Class ____________

LESSON 10-6

Reteach

Volume of Prisms and Cylinders

Volume of Prisms	
Prism	The volume of a prism with base area B and height h is $V = Bh$.
Right Rectangular Prism	The volume of a right rectangular prism with length ℓ, width w, and height h is $V = \ell wh$.
Cube	The volume of a cube with edge length s is $V = s^3$.

Volume of a Cylinder

The volume of a cylinder with base area B, radius r, and height h is

$$V = Bh, \text{ or } V = \pi r^2 h.$$

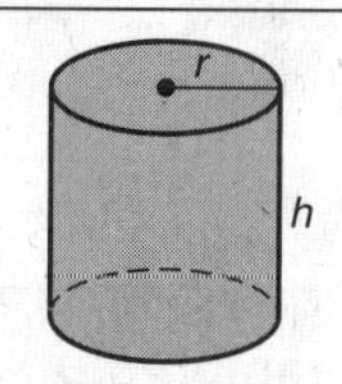

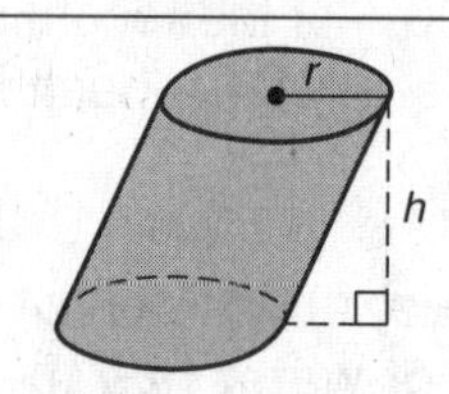

Find the volume of each prism.

1.

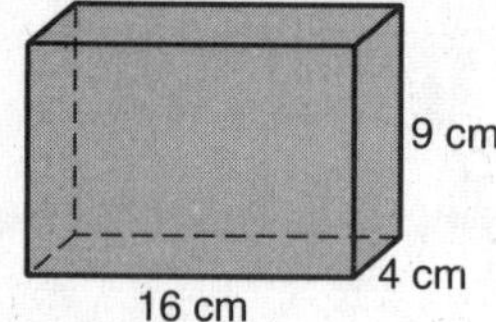

2.

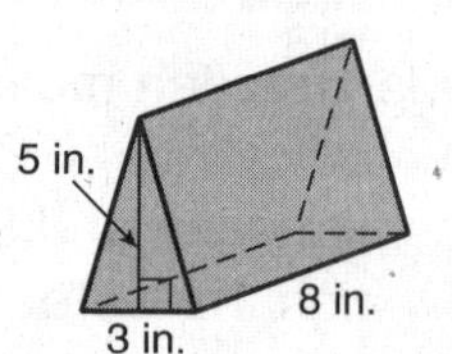

Find the volume of each cylinder. Give your answers both in terms of π and rounded to the nearest tenth.

3.

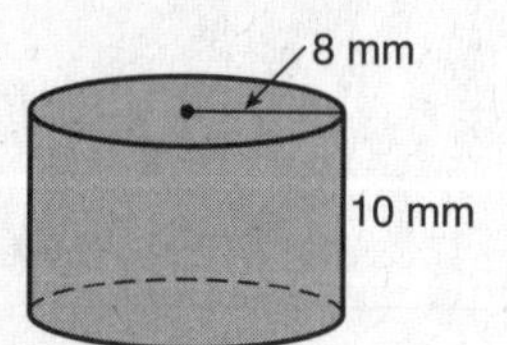

4.

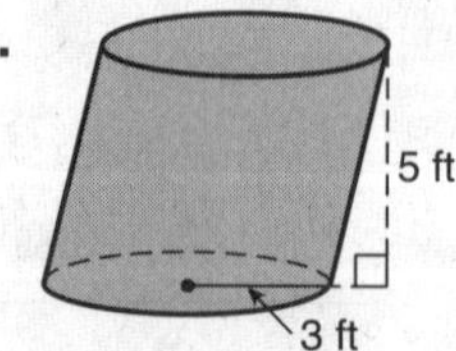

Copyright © by Holt, Rinehart and Winston.
All rights reserved.

Name ______________________ Date __________ Class __________

LESSON 10-6

Reteach

Volume of Prisms and Cylinders continued

The dimensions of the prism are multiplied by $\frac{1}{3}$. Describe the effect on the volume.

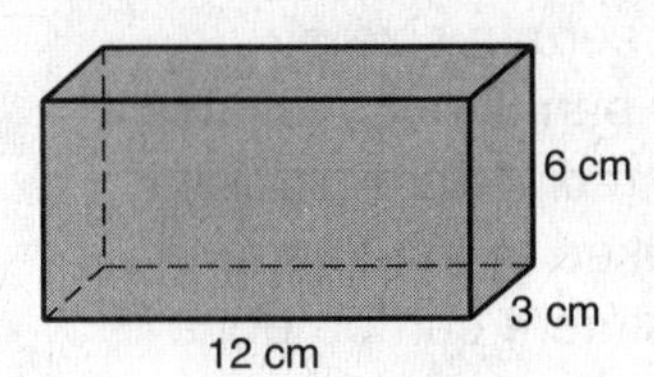

original volume:

$V = \ell wh$

$= (12)(3)(6)$ $\quad \ell = 12, w = 3, h = 6$

$= 216 \text{ cm}^3$ $\quad$ Simplify.

new volume, dimensions multiplied by $\frac{1}{3}$:

$V = \ell wh$

$= (4)(1)(2)$ $\quad \ell = 4, w = 1, h = 2$

$= 8 \text{ cm}^3$ $\quad$ Simplify.

Notice that $216 \cdot \frac{1}{27} = 8$. If the dimensions are multiplied by $\frac{1}{3}$, the volume is multiplied by $\left(\frac{1}{3}\right)^3$, or $\frac{1}{27}$.

Describe the effect of each change on the volume of the given figure.

5. The dimensions are multiplied by 2.

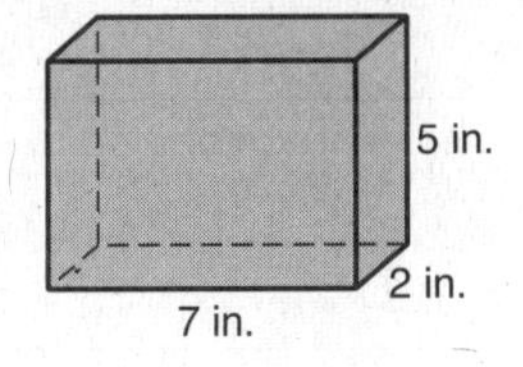

____

6. The dimensions are multiplied by $\frac{1}{4}$.

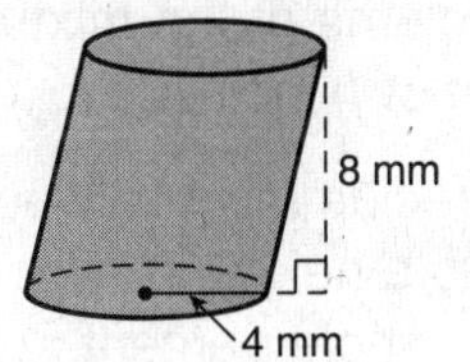

____

Find the volume of each composite figure. Round to the nearest tenth.

7.

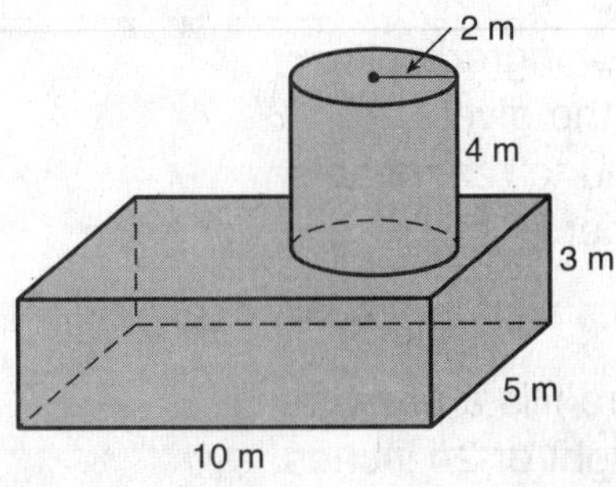

____

8.

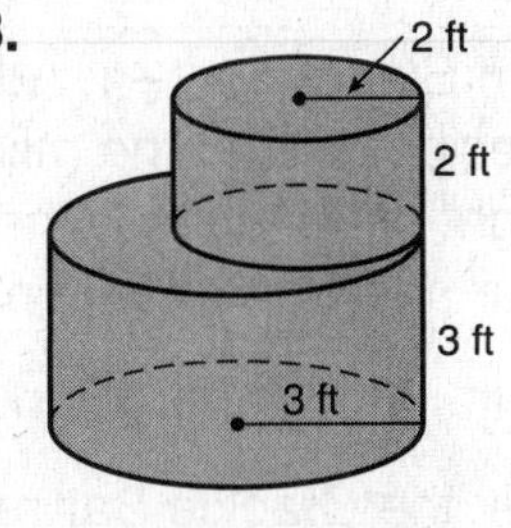

____

Copyright © by Holt, Rinehart and Winston.
All rights reserved.

Name ______________________ Date __________ Class __________

LESSON 10-6

Challenge

Using the Volume Formula to Adjust a Recipe

Most baking recipes specify a certain size of baking pan. When that size of pan is not available, you may be able to adjust the recipe to a different size. Since many items are baked in the shape of a rectangular prism, this adjustment can be done by calculating volumes with the formula $V = \ell wh$. For example, the recipe at right requires a $13 \times 9 \times 2$-inch pan. This type of pan is shaped like a rectangular prism that is 13 inches long, 9 inches wide, and 2 inches high, as shown below.

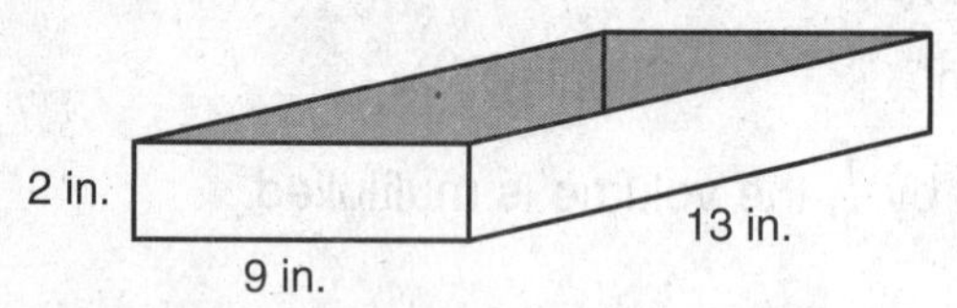

Crackle Bars

3 tablespoons margarine
40 regular marshmallows, or
4 cups miniature marshmallows
6 cups toasted rice cereal

Melt the margarine in a large saucepan over low heat. Add the marshmallows and stir until they are completely melted. Remove from heat. Stir in the cereal until it is coated. Press the mixture into a greased 13 × 9 × 2-inch pan. Cut bars when cool.

Refer to the recipe for Crackle Bars that is given above. Assume that when you prepare the mixture according to the recipe, it fills the 13 × 9 × 2-inch pan to an unknown height of *h* inches.

1. What is the volume of the recipe mixture? ______________

2. Suppose that you only have an $8 \times 8 \times 2$-inch pan. What would be the volume of the mixture if the pan were filled to a height of h inches? ______________

3. What percent of the recipe mixture would fill the $8 \times 8 \times 2$-inch pan to a height of h inches? Round to the nearest whole percent. ______________

4. Calculate the amount of each ingredient needed to make enough mixture to fill the $8 \times 8 \times 2$-inch pan to a height of h inches with no extra mixture.

a. margarine (1 tablespoon equals 3 teaspoons) ______________

b. regular marshmallows ______________

c. miniature marshmallows ______________

d. toasted rice cereal ______________

5. Use the method from Exercises 1–4. Adjust the amounts of ingredients in the Crackle Bar recipe so that the mixture fills a pan of the given dimensions to a height of h inches. When necessary, round to reasonable measures. Write your answers on a separate sheet of paper.

a. 9 in. × 9 in. × 2 in. **b.** 15 in. × 10 in. × $1\frac{1}{2}$ in. **c.** 25 cm × 35 cm × 4 cm

6. Explain how to adjust the Crackle Bar recipe so the mixture fills a pan that is a inches long, b inches wide, and c inches high to a height of $2h$ inches.

____

____

Copyright © by Holt, Rinehart and Winston.
All rights reserved.

Name ______________________ Date __________ Class __________

LESSON 10-6

Problem Solving

Volume of Prisms and Cylinders

1. A cylindrical juice container has the dimensions shown. About how many cups of juice does this container hold? (*Hint:* 1 cup $\approx$ 14.44 in^3)

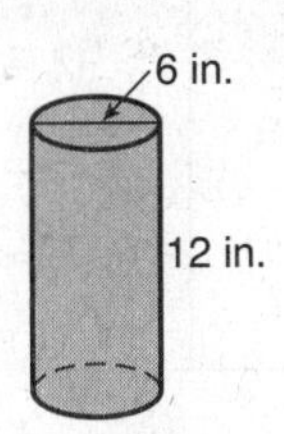

____

2. A large cylindrical cooler is $2\frac{1}{2}$ feet high and has a diameter of $1\frac{1}{2}$ feet. It is filled $\frac{3}{4}$ high with water for athletes to use during their soccer game. Estimate the volume of the water in the cooler in gallons. (*Hint:* 1 gallon $\approx$ 231 in^3)

____

Choose the best answer.

3. How many 3-inch cubes can be placed inside the box?

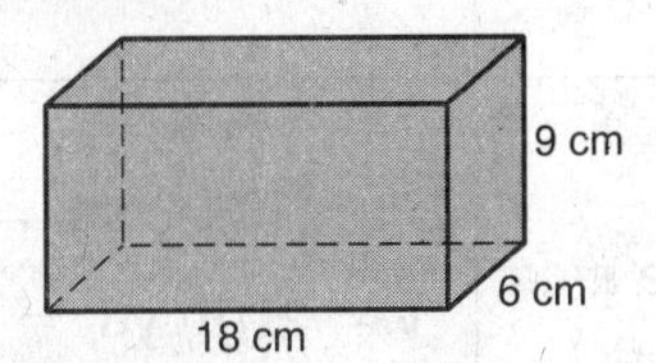

A 27   C 45

B 36   D 72

4. A cylinder has a volume of 4π cm^3. If the radius and height are each tripled, what will be the new volume of the cylinder?

F 12π cm^3 H 64π cm^3

G 36π cm^3 J 108π cm^3

5. What is the volume of the composite figure with the dimensions shown in the three views? Round to the nearest tenth.

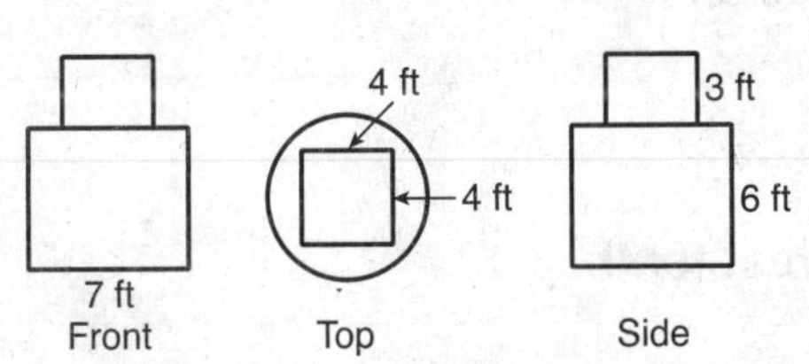

A 182.9 ft^3 C 278.9 ft^3

B 205.7 ft^3 D 971.6 ft^3

6. Find the expression that can be used to determine the volume of the composite figure shown.

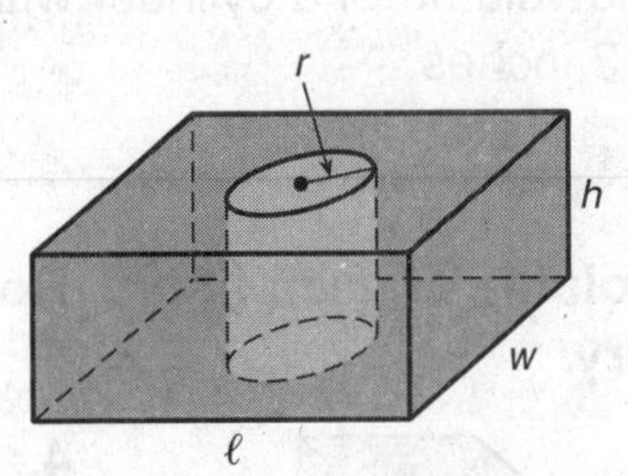

F $\ell wh - \pi r^2 h$ H $\pi r^2 h - \ell wh$

G $\pi r^2 h + \ell wh$ J $\ell wh + 2\pi r^2 h$

Copyright © by Holt, Rinehart and Winston.
All rights reserved.

Name ______________________ Date __________ Class __________

LESSON 10-6

Reading Strategies

Use a Table

The tables below show formulas for finding the volume of different three-dimensional figures.

Cylinders	**Diagram**	**Formula**	**Example**
Oblique and right cylinders	4 cm; 3 cm	$V = Bh$ where B is the area of the base, so $V = \pi r^2 h$	$V = \pi(4^2)(3)\text{ cm}^3$ $= \pi(48)\text{ cm}^3$ $\approx 150.8\text{ cm}^3$

Prisms	**Diagram**	**Formula**	**Example**
Cube	13 m	$V = s^3$	$V = 13^3\text{ m}^3$ $= 2197\text{ m}^3$
Rectangular prism	8 ft; 5 ft; 11 ft	$V = \ell wh$	$V = (11)(5)(8)\text{ ft}^3$ $= 440\text{ ft}^3$
Other oblique and right prisms	7 yd; 8 yd; 6 yd	$V = Bh$ where B is the area of the base	$B = \frac{1}{2}(6)(7)\text{ yd}^2 = 21\text{ yd}^2$ $V = 21(8)\text{ yd}^3 = 168\text{ yd}^3$

Answer the following. Round to the nearest tenth if necessary.

1. Find the volume of a cube with edge length 8 centimters. ______________

2. Find the volume of a cylinder with height 15 inches and radius 3 inches. ______________

Find the volume of each figure. Round to the nearest tenth if necessary.

3.

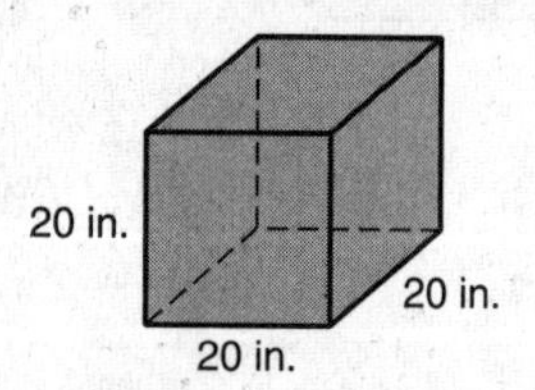

____

4.

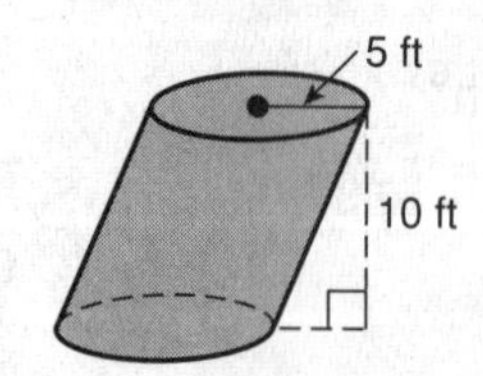

____

5.

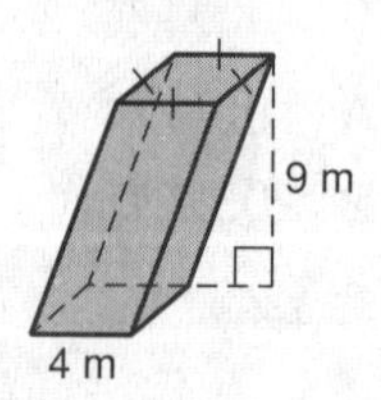

____

Copyright © by Holt, Rinehart and Winston.
All rights reserved.

Name ______________________ Date __________ Class __________

LESSON 10-7

Practice A

Volume of Pyramids and Cones

Write each formula.

1. volume of a pyramid with base area B and height h ______________

2. volume of a cone with radius r and height h ______________

Find the volume of each pyramid.

3. 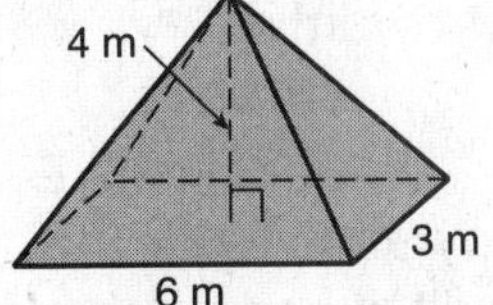

the rectangular pyramid

____

4. 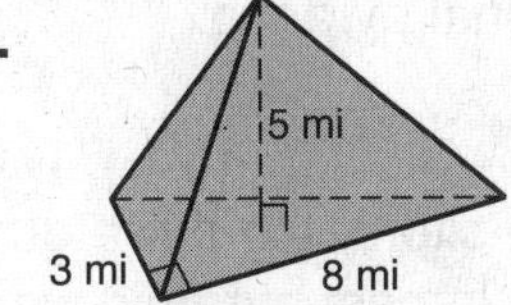

the right triangular pyramid

____

5. a square pyramid with side length 10 in. and height 12 in. ______________

Find the volume of each cone. Give your answers both in terms of π and rounded to the nearest tenth.

6. 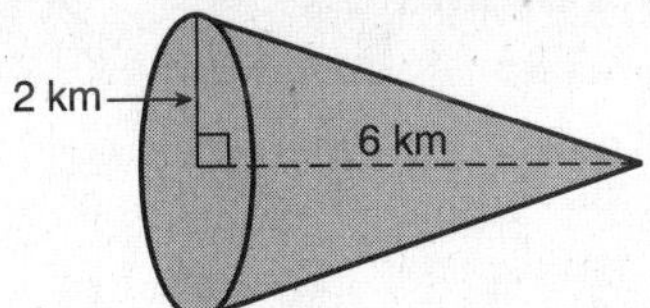

____

7. a cone with diameter 15 yd and height 10 yd

____

8. An ant lion is an insect that digs cone-shaped pits in loose dirt to trap ants. When an ant tumbles down into the pit, the ant lion eats it. A typical ant lion pit has a radius of 1 inch and a depth of 2 inches. Find the volume of dirt the ant lion moved to dig its hole. Round to the nearest tenth.

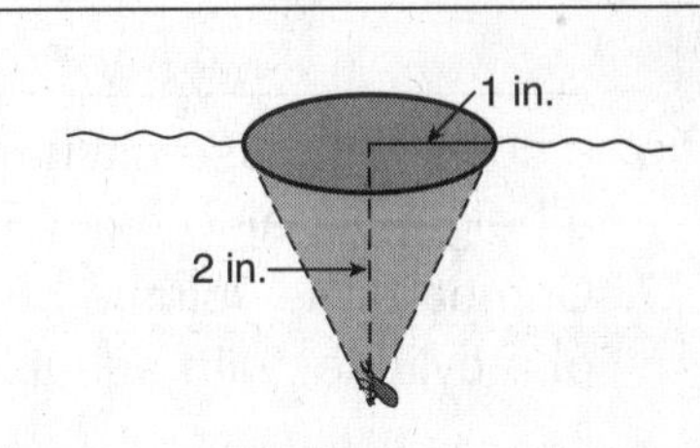

____

Complete Exercises 9–11 to describe the effect on the volume of dividing the dimensions of the cone by 3. Give your answers in terms of π.

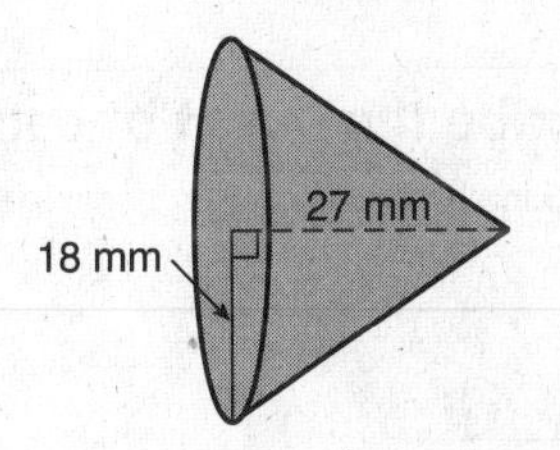

9. Find the volume of the cone. ______________

10. Find the volume of the cone after the radius and height are divided by 3. ______________

11. Describe the effect on the volume after dividing the dimensions of a cone by 3. ______________

12. Find the volume of the composite figure.

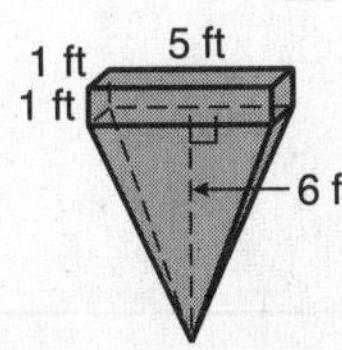

____

Copyright © by Holt, Rinehart and Winston.
All rights reserved.

Name ______________________________ Date ____________ Class ____________

LESSON **10-7**

Practice B

Volume of Pyramids and Cones

Find the volume of each pyramid. Round to the nearest tenth if necessary.

1.

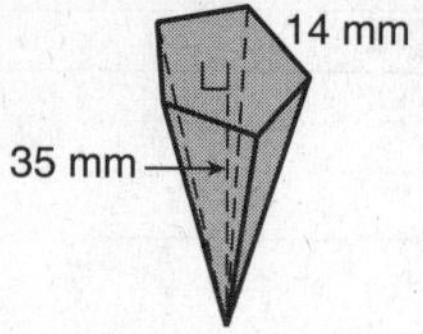

the regular pentagonal pyramid

____

2.

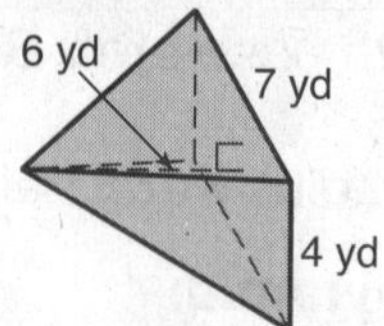

the rectangular right pyramid

____

3. Giza in Egypt is the site of the three great Egyptian pyramids. Each pyramid has a square base. The largest pyramid was built for Khufu. When first built, it had base edges of 754 feet and a height of 481 feet. Over the centuries, some of the stone eroded away and some was taken for newer buildings. Khufu's pyramid today has base edges of 745 feet and a height of 471 feet. To the nearest cubic foot, find the difference between the original and current volumes of the pyramid.

____

Find the volume of each cone. Give your answers both in terms of π and rounded to the nearest tenth.

4.

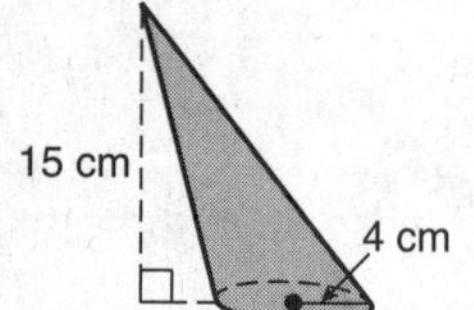

____

5.

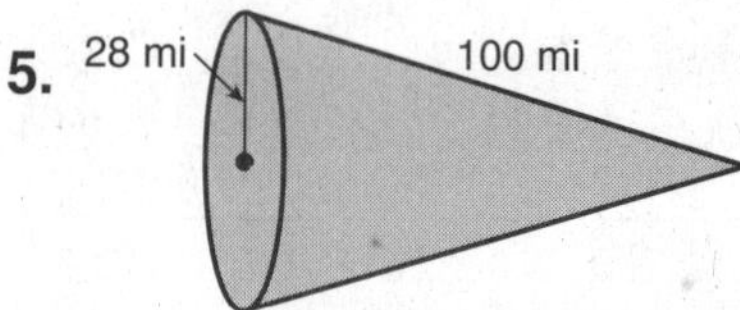

____

6. a cone with base circumference 6π m and a height equal to half the radius

____

7. Compare the volume of a cone and the volume of a cylinder with equal height and base area.

____

Describe the effect of each change on the volume of the given figure.

8.

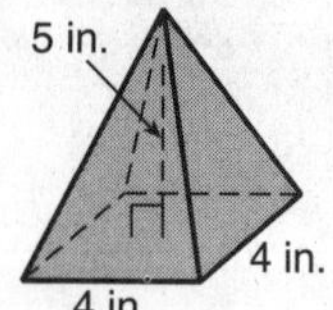

The dimensions are multiplied by $\frac{2}{3}$.

____

9.

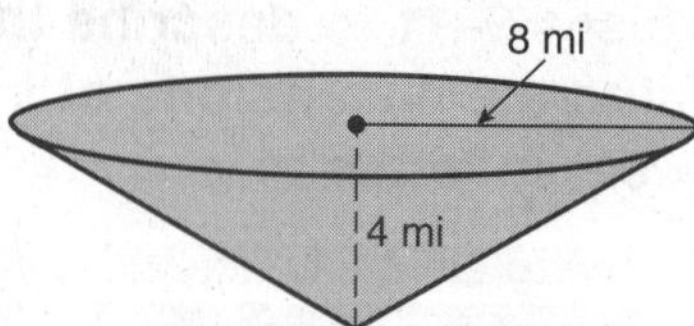

The dimensions are tripled.

____

Find the volume of each composite figure. Round to the nearest tenth.

10.

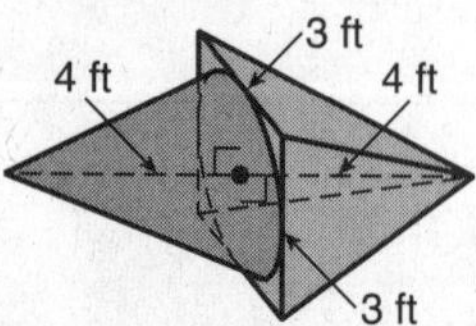

____

11.

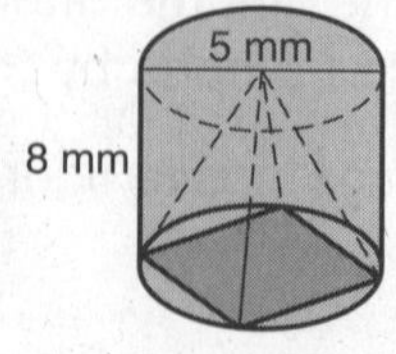

____

Copyright © by Holt, Rinehart and Winston.
All rights reserved.

Name ______________________ Date __________ Class __________

LESSON 10-7

Practice C
Volume of Pyramids and Cones

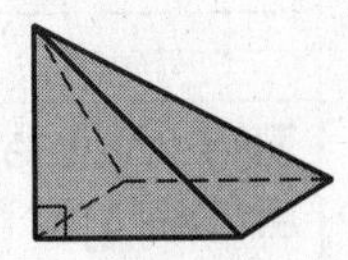

1. The figure shows a square-based pyramid with a height equal to the side length of the base. The segment connecting the vertex to its closest corner is perpendicular to the base. Draw a net for this pyramid below. Then, using a separate sheet of paper, cut out three of these shapes. Fold them into pyramids, and assemble them into a cube. Describe what this demonstrates.

____

____

2. A square pyramid has a height equal to its base's side length, and its surface area is equal to its volume (although the units are different). Find the side length of the base. Give both an exact answer and an answer rounded to the nearest tenth. ____________

3. A cone has a height equal to its radius, and its surface area is equal to its volume (although the units are different). Find the radius. Give both an exact answer and an answer rounded to the nearest tenth. ____________

4. Draw a figure that has exactly two-thirds the volume of this regular hexagonal prism.

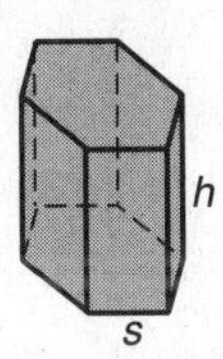

Find the volume of each figure. Round to the nearest tenth if necessary.

5. 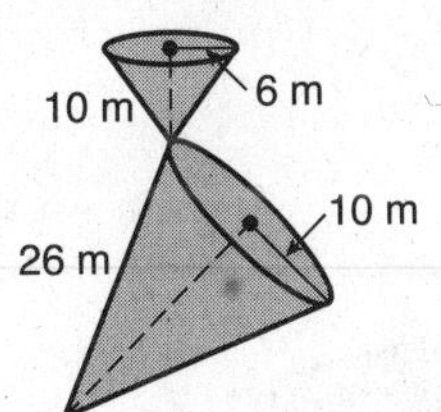

____

6. 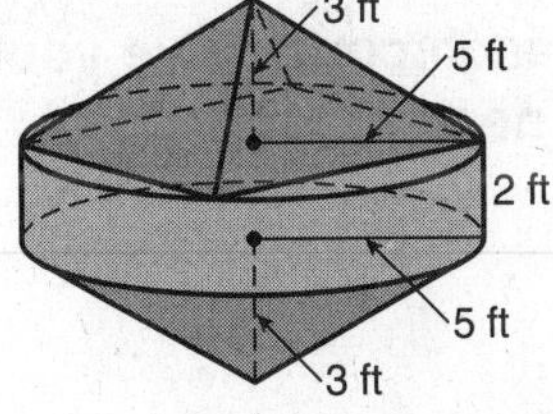

____

7. 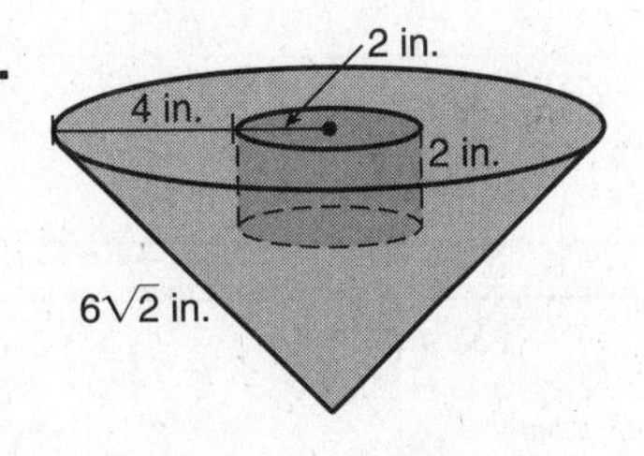

____

8. 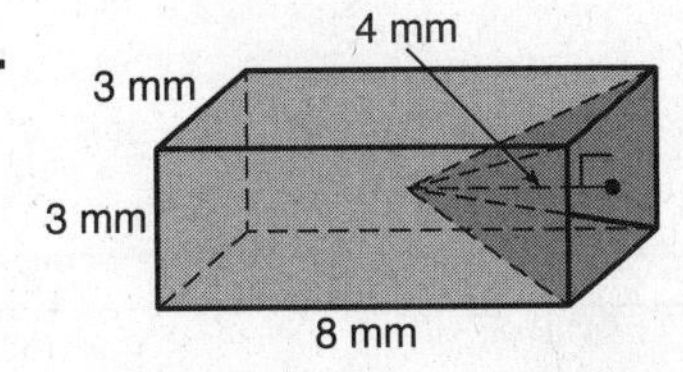

____

Copyright © by Holt, Rinehart and Winston.
All rights reserved.

Name ______________________ Date __________ Class __________

LESSON 10-7

Reteach

Volume of Pyramids and Cones

Volume of a Pyramid
The volume of a pyramid with base area B and height h is $V = \frac{1}{3}Bh$.

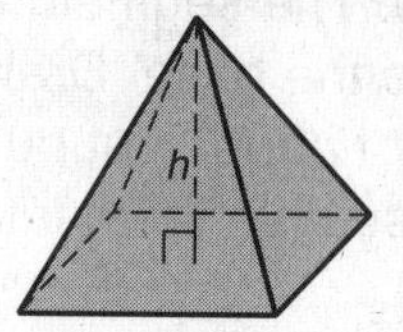

Volume of a Cone
The volume of a cone with base area B, radius r, and height h is $V = \frac{1}{3}Bh$, or $V = \frac{1}{3}\pi r^2 h$.

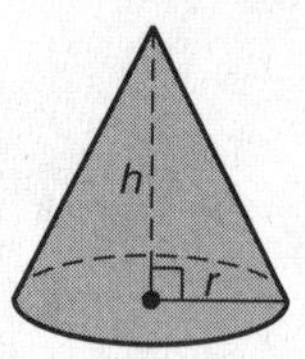

Find the volume of each pyramid. Round to the nearest tenth if necessary.

1\.

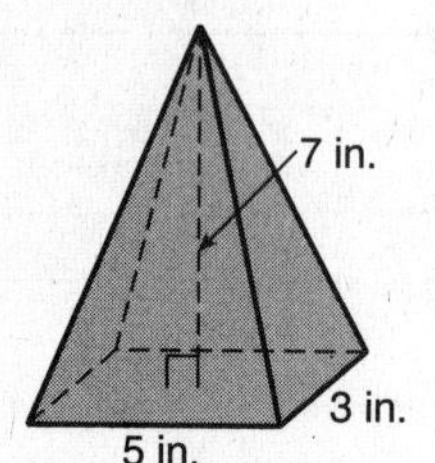

____

2\.

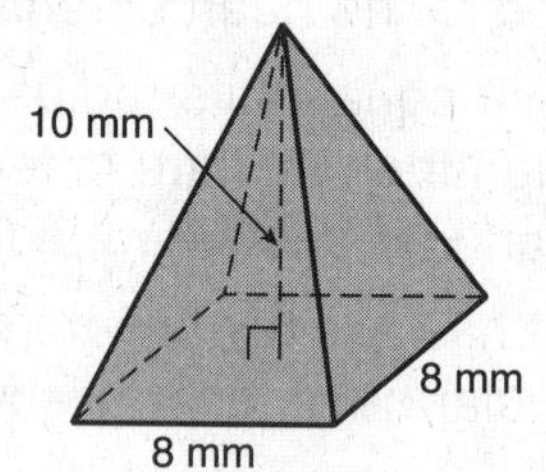

____

Find the volume of each cone. Give your answers both in terms of π and rounded to the nearest tenth.

3\.

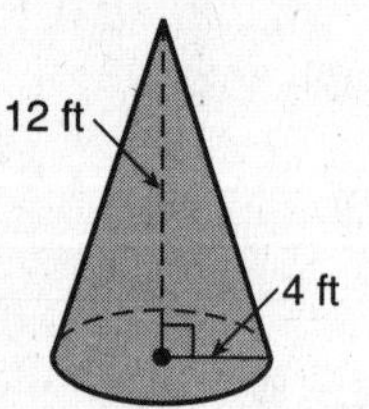

____

4\.

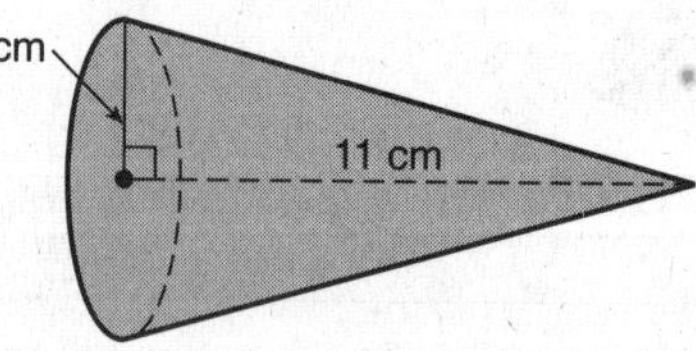

____

Copyright © by Holt, Rinehart and Winston.
All rights reserved.

Name ______________________________ Date ____________ Class ____________

LESSON 10-7

Reteach

Volume of Pyramids and Cones *continued*

The radius and height of the cone are multiplied by $\frac{1}{2}$. Describe the effect on the volume.

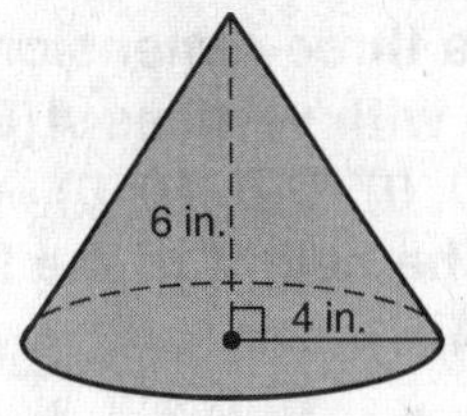

original volume:

$V = \frac{1}{3}\pi r^2 h$

$= \frac{1}{3}\pi(4)^2(6)$ $\quad r = 4,\ h = 6$

$= 32\pi \text{ in}^3$ $\quad$ Simplify.

new volume, dimensions multiplied by $\frac{1}{2}$:

$V = \frac{1}{3}\pi r^2 h$

$= \frac{1}{3}\pi(2)^2(3)$ $\quad r = 2,\ h = 3$

$= 4\pi \text{ in}^3$ $\quad$ Simplify.

If the dimensions are multiplied by $\frac{1}{2}$, then the volume is multiplied by $\left(\frac{1}{2}\right)^3$, or $\frac{1}{8}$.

Describe the effect of each change on the volume of the given figure.

5. The dimensions are doubled.

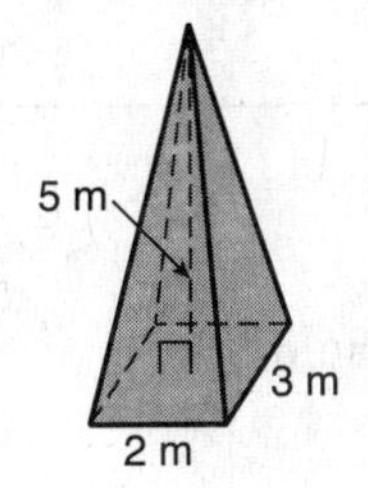

____

6. The radius and height are multiplied by $\frac{1}{3}$.

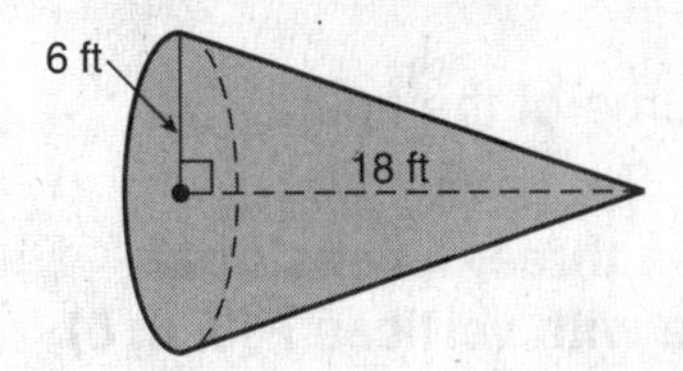

____

Find the volume of each composite figure. Round to the nearest tenth if necessary.

7.

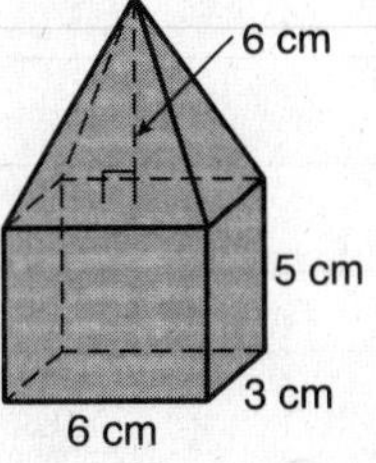

____

8.

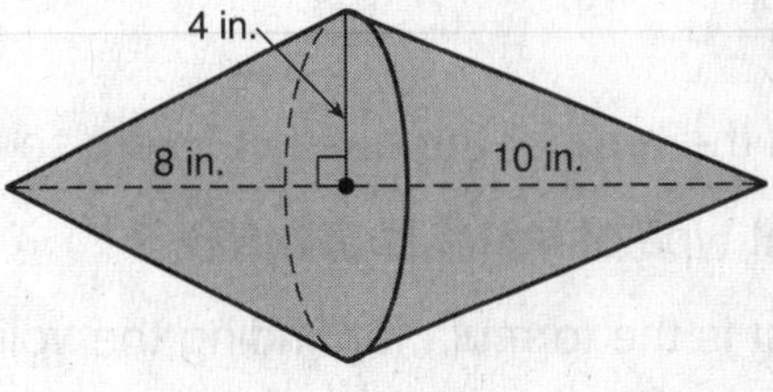

____

Copyright © by Holt, Rinehart and Winston.
All rights reserved.

Name ______________________ Date ____________ Class ____________

LESSON 10-7

Challenge

Volume of Pyramids and Cones

Draw a figure in a three-dimensional coordinate plane with vertices $A(0, 0, 0)$, $B(8, 0, 0)$, $C(0, 10, 0)$, $D(8, 10, 0)$, and $E(4, 4, 12)$. The height of the figure is EF. F is at $(4, 4, 0)$.

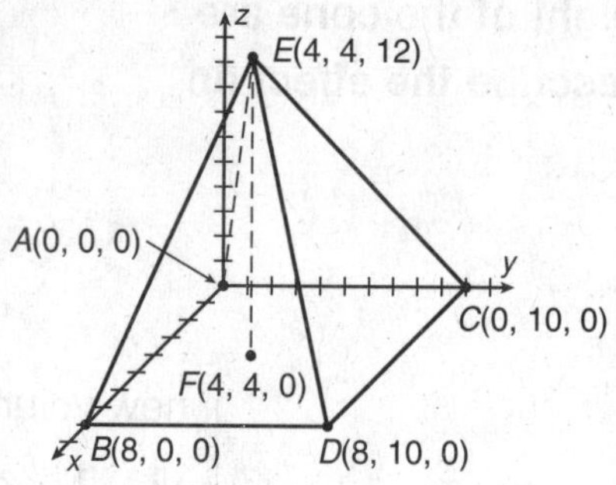

1. Name the base of this figure and its shape. ______________

2. What type of figure is *ABDCE*? ______________

3. What is the formula for finding the volume of this figure? ______________

4. Use the distance formula to find *EF*. ______________

5. Find *AB*. ______________

6. Find *AC*. ______________

7. Find the volume of the figure. ______________

Draw a figure in a three dimensional coordinate plane with vertices $P(0, 0, 0)$, $N(-10, 0, 0)$, $L(0, -10, 0)$, $M(-10, -10, 0)$, $K(-5, -5, -6)$, $J(-5, -5, 7)$.

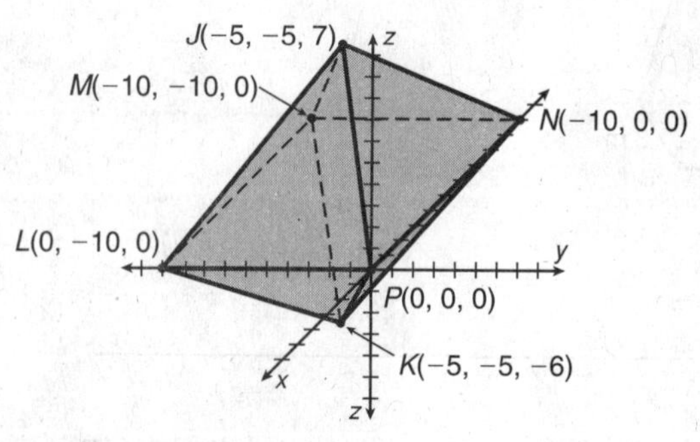

8. Name the base of this figure and its shape. You might want to plot the base coordinates on an *x-y* plane. ______________

9. Find *LP*. ______________

10. Find the area of the base in Exercise 8. ______________

11. What type of figure is *JLMNPK*? ______________

12. What is the formula for finding the volume of this figure? Explain.

____

____

13. Find the volume of the three-dimensional shape named in Exercise 11. Round to the nearest tenth. ______________

Copyright © by Holt, Rinehart and Winston. All rights reserved.

Name ______________________ Date __________ Class __________

LESSON 10-7

Problem Solving

Volume of Pyramids and Cones

1. A regular square pyramid has a base area of 196 meters and a lateral area of 448 square meters. What is the volume of the pyramid? Round your answer to the nearest tenth.

____

2. A paper cone for serving roasted almonds has a volume of 406π cubic centimeters. A smaller cone has half the radius and half the height of the first cone. What is the volume of the smaller cone? Give your answer in terms of π.

____

3. The hexagonal base in the pyramid is a regular polygon. What is the volume of the pyramid if its height is 9 centimeters? Round to the nearest tenth.

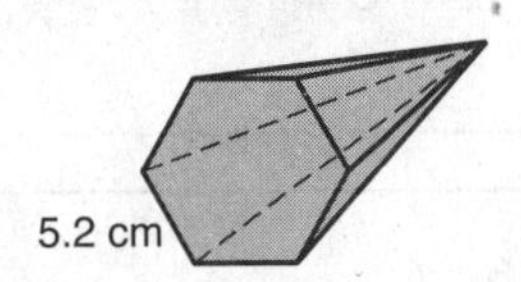

____

4. Find the volume of the shaded solid in the figure shown. Give your answer in terms of π.

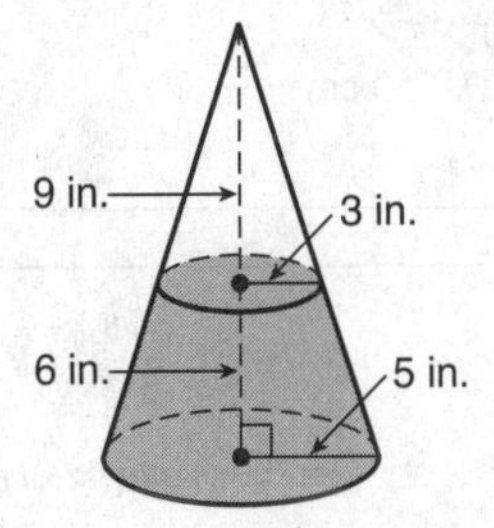

____

Choose the best answer.

5. The diameter of the cone equals the width of the cube, and the figures have the same height. Find the expression that can be used to determine the volume of the composite figure.

A $4(4)(4) - \frac{1}{3}\pi(2^2)(4)$

B $4(4)(4) + \frac{1}{3}\pi(2^2)(4)$

C $4(4)(4) - \pi(2^2)(4)$

D $4(4)(4) + \frac{1}{3}\pi(2^2)$

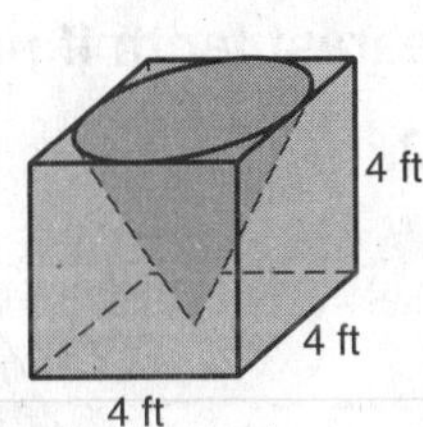

6. Approximately how many fluid ounces of water can the paper cup hold? (*Hint:* 1 fl oz $\approx$ 1.805 in^3)

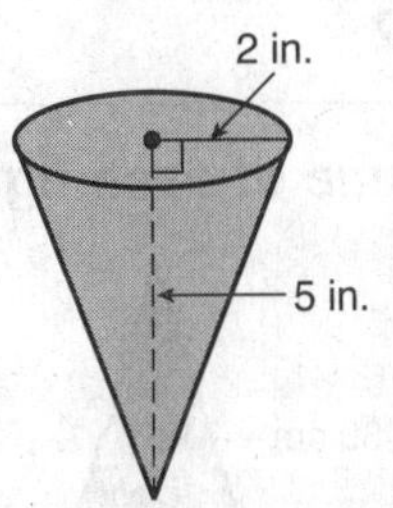

F 10.9 fl oz **H** 32.7 fl oz

G 11.6 fl oz **J** 36.3 fl oz

7. The Step Pyramid of Djoser in Lower Egypt was the first pyramid in the history of architecture. Its original height was 204 feet, and it had a rectangular base measuring 411 feet by 358 feet. Which is the best estimate for the volume of the pyramid in cubic yards?

A 370,570 yd^3 **C** 3,335,128 yd^3

B 1,111,709 yd^3 **D** 10,005,384 yd^3

Copyright © by Holt, Rinehart and Winston.
All rights reserved.

Name ______________________ Date ____________ Class ____________

LESSON 10-7

Reading Strategies

Use a Concept Map

Use the concept maps below to help you understand and use formulas for volume.

Formula

$$V = \frac{1}{3}Bh$$

h = height of pyramid and B = area of base

Volume of Pyramids

Diagram

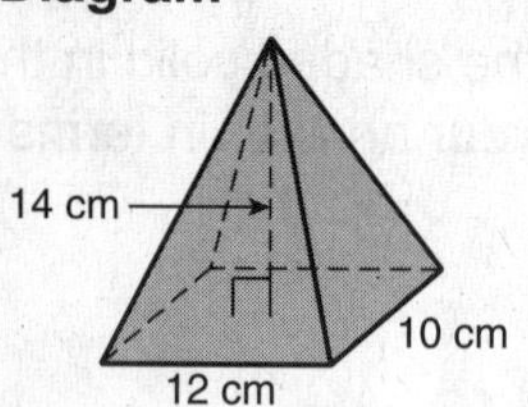

Example

$V = \frac{1}{3}Bh$

$B = 12(10)\text{ cm}^2 = 120\text{ cm}^2$

$V = \frac{1}{3}(120)(14)\text{ cm}^3 = 560\text{ cm}^3$

Formulas

$$V = \frac{1}{3}Bh \quad \text{or} \quad V = \frac{1}{3}\pi r^2 h$$

r = radius, h = height of cone, and B = area of base

Volume of Cones

Diagram

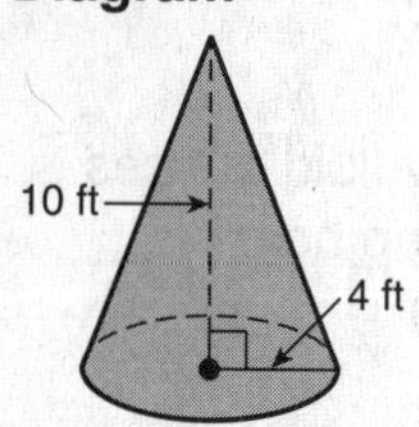

Example

$V = \frac{1}{3}Bh$

$B = \pi(4)^2\text{ ft}^2 = 16\pi\text{ ft}^2$

$V = \frac{1}{3}(16\pi)(10)\text{ ft}^3 \approx 167.6\text{ ft}^3$

Find the volume of each figure. Round to the nearest tenth if necessary.

1.

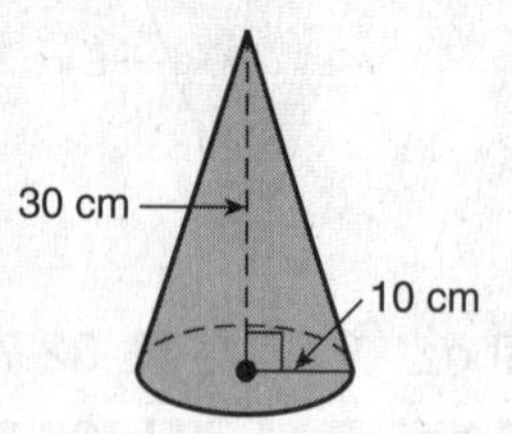

____

2.

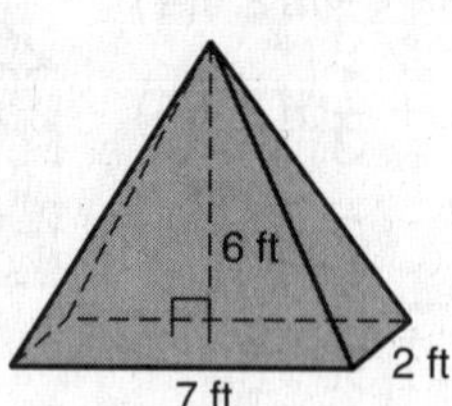

____

3.

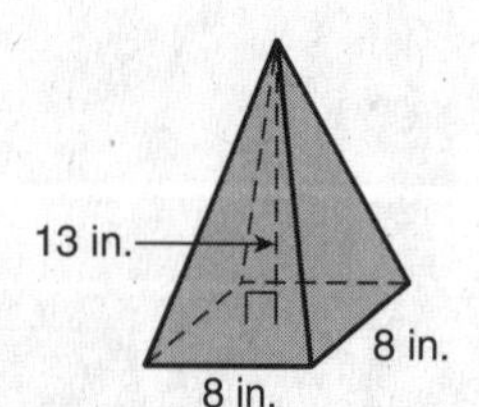

____

4.

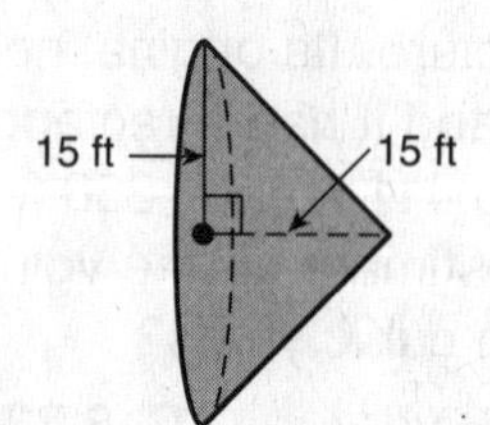

____

Copyright © by Holt, Rinehart and Winston.
All rights reserved.

Name ______________________ Date ____________ Class ____________

LESSON **10-8**

Practice A

Spheres

Write each formula.

1. volume of a sphere with radius r ____________

2. surface area of a sphere with radius r ____________

Find each measurement. Give your answers in terms of π.

3.

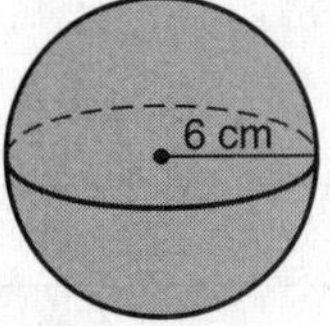

the volume of the sphere

____

4.

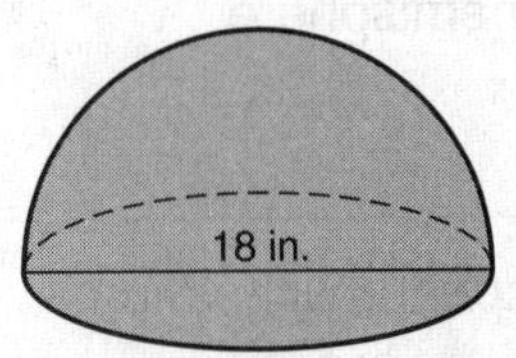

the volume of the hemisphere

____

5. the radius of a sphere with a volume of $36{,}000\pi$ mm^3 ____________

6. Margot is thirsty after a 5-km run for charity. The organizers offer the containers of water shown in the figure. Margot wants the one with the greater volume of water. Tell which container Margot should pick.

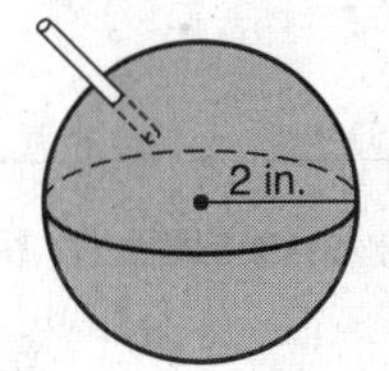

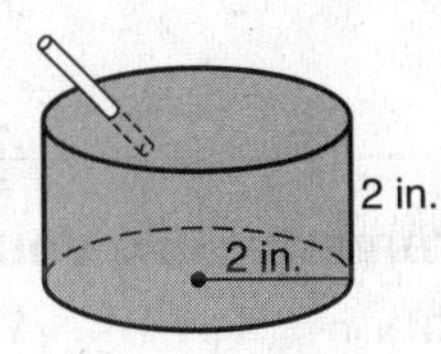

____

Find the surface area of each sphere. Give your answers in terms of π.

7.

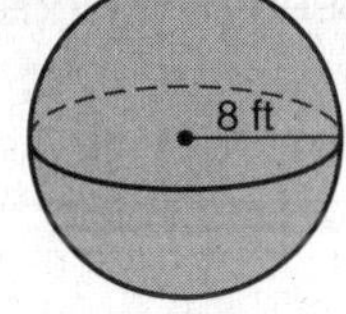

____

8. the surface area of a sphere with volume $\frac{256\pi}{3}$ yd^3

____

Complete Exercises 9–11 to describe the effect on the volume and the surface area of multiplying the radius of a sphere by 3.

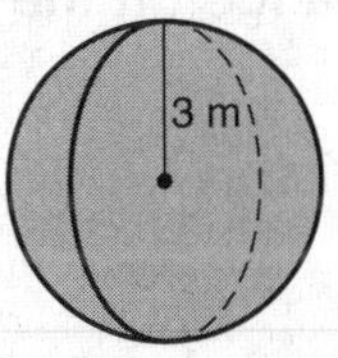

9. Find the volume and surface area of the sphere. ____________

10. Find the volume and surface area of the sphere after the radius is multiplied by 3. ____________

11. Describe the effect on the volume and surface area of multiplying the radius of the sphere by 3.

____

12. Find the volume and surface area of the composite figure. Give your answers in terms of π.

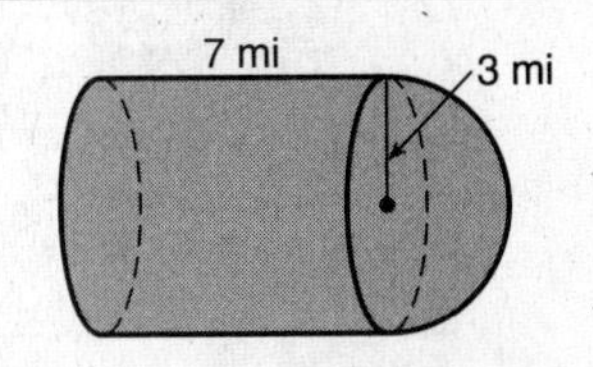

____

Copyright © by Holt, Rinehart and Winston.
All rights reserved.

Name ______________________ Date __________ Class __________

LESSON 10-8

Practice B
Spheres

Find each measurement. Give your answers in terms of π.

1. 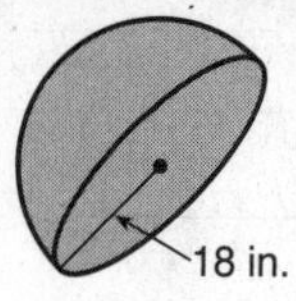

the volume of the hemisphere

____

2. 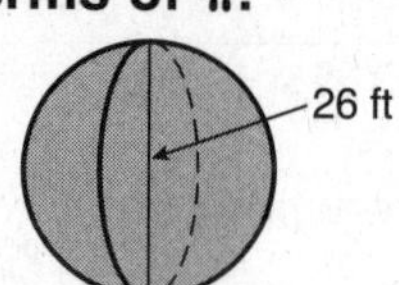

the volume of the sphere

____

3. the diameter of a sphere with volume $\frac{500\pi}{3}$ m^3 ________________________

4. The figure shows a grapefruit half. The radius to the outside of the rind is 5 cm. The radius to the inside of the rind is 4 cm. The edible part of the grapefruit is divided into 12 equal sections. Find the volume of the half grapefruit and the volume of one edible section. Give your answers in terms of π.

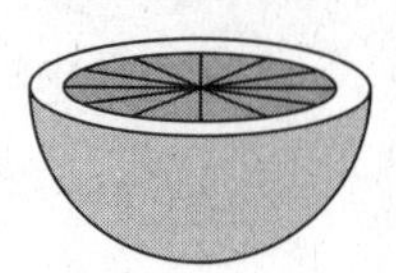

____

Find each measurement. Give your answers in terms of π.

5. 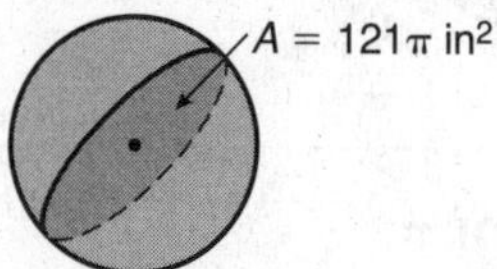

the surface area of the sphere

____

6. 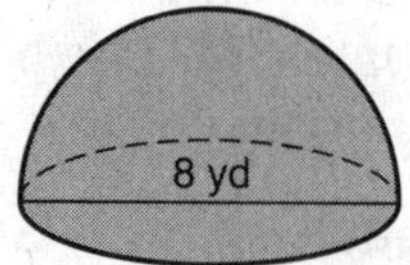

the surface area of the closed hemisphere and its circular base

____

7. the volume of a sphere with surface area 196π km^2

____

Describe the effect of each change on the given measurement of the figure.

8. 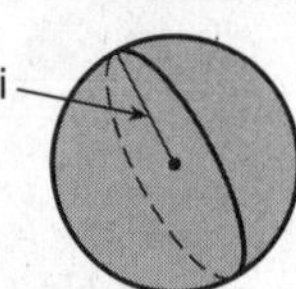

surface area
The dimensions are divided by 4.

____

9. 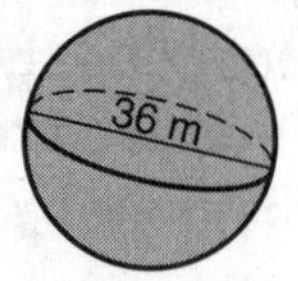

volume
The dimensions are multiplied by $\frac{2}{5}$.

____

Find the surface area and volume of each composite figure. Round to the nearest tenth.

10. 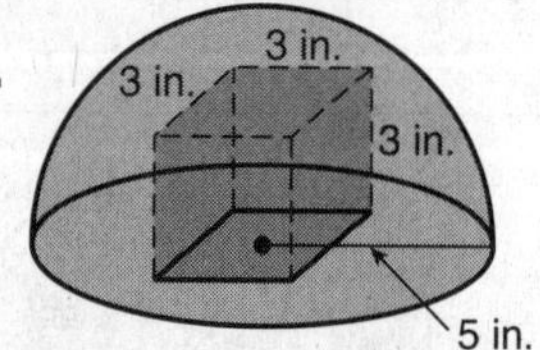

____

11. 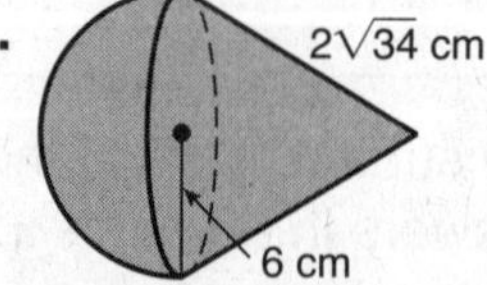

____

Copyright © by Holt, Rinehart and Winston.
All rights reserved.

Name ________________________ Date __________ Class __________

LESSON 10-8

Practice C
Spheres

1. A sphere has radius r. Draw a composite figure made up of a square prism (not a cube) and a square pyramid that has the same volume as the sphere.

2. Find the surface area of the composite figure you drew in Exercise 1.

____

3. Consider a composite figure made up of a cylinder and a cone that has the same volume as a sphere with radius r. Find the figure's surface area. ____________

Use the figure for Exercises 4–6. The figure shows a hollow, sealed container with some water inside.

4. There is just enough water in the container to exactly fill the hemisphere. The container is held so that the point of the cone is down and the altitude of the cone is exactly vertical. Find the height of the water in the cone. Round to the nearest tenth. ____________

5. Suppose the amount of water in the container is exactly enough to fill the cone. The container is held so that the hemisphere is down and the altitude of the cone is exactly vertical. Find the height of the water in the container. Round to the nearest tenth. ____________

6. Find the height of the cone with the same radius if the container were made so that the water would exactly fill either the hemisphere or the cone. ____________

7. A sphere has center (0, 0, 0). Its surface passes through the point (x, y, z). Find the sphere's surface area and volume.

____

Use the figure for Exercises 8–10. The figure shows a can of three tennis balls. The can is just large enough so that the tennis balls will fit inside with the lid on. The diameter of each tennis ball is 2.5 in. Give exact fraction answers.

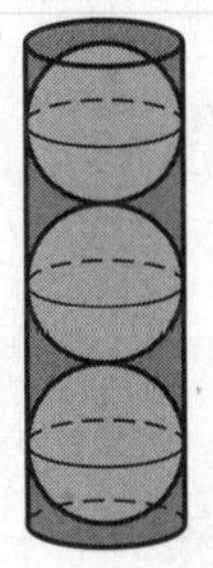

8. Find the total volume of the can. ____________

9. Find the volume of empty space inside the can. ____________

10. Tell what percent of the can is occupied by the tennis balls. ____________

Copyright © by Holt, Rinehart and Winston.
All rights reserved.

Name ______________________ Date ____________ Class ____________

LESSON **10-8**

Reteach

Spheres

Volume and Surface Area of a Sphere		
Volume	The volume of a sphere with radius r is $V = \frac{4}{3}\pi r^3$.	
Surface Area	The surface area of a sphere with radius r is $S = 4\pi r^2$.	

Find each measurement. Give your answer in terms of π.

1. the volume of the sphere

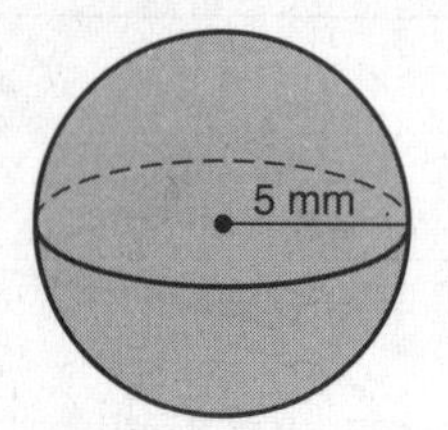

____

2. the volume of the sphere

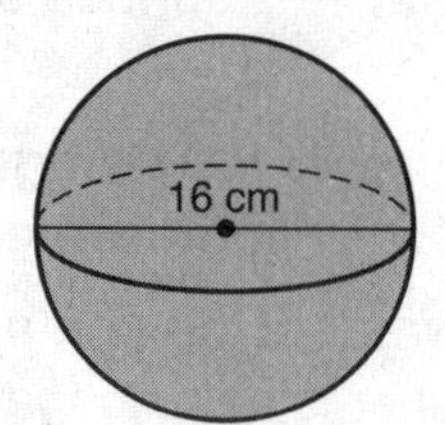

____

3. the volume of the hemisphere

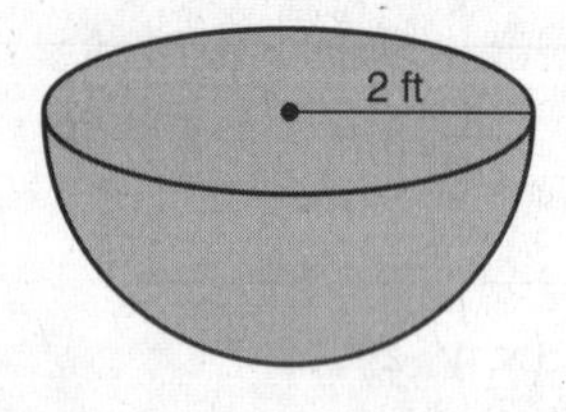

____

4. the radius of a sphere with volume 7776π in^3

____

5. the surface area of the sphere

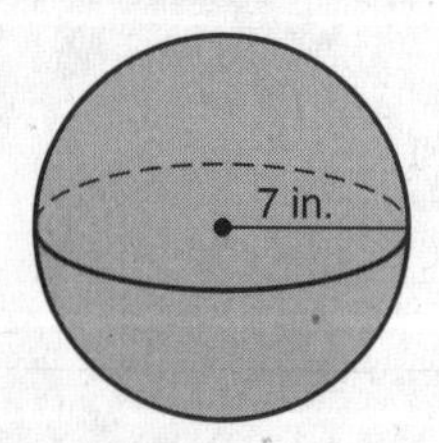

____

6. the surface area of the sphere

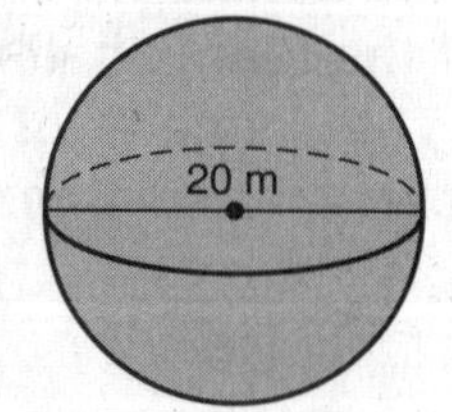

____

Copyright © by Holt, Rinehart and Winston.
All rights reserved.

Name ______________________ Date __________ Class __________

LESSON 10-8

Reteach

Spheres continued

The radius of the sphere is multiplied by $\frac{1}{4}$. Describe the effect on the surface area.

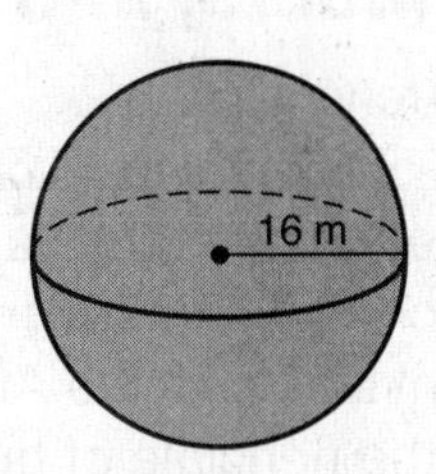

original surface area:

$S = 4\pi r^2$

$= 4\pi(16)^2$ $r = 16$

$= 1024\pi \text{ m}^2$ Simplify.

new surface area, radius multiplied by $\frac{1}{4}$:

$S = 4\pi r^2$

$= 4\pi(4)^2$ $r = 4$

$= 64\pi \text{ m}^2$ Simplify.

Notice that $1024 \cdot \frac{1}{16} = 64$. If the dimensions are multiplied by $\frac{1}{4}$, the surface area is multiplied by $\left(\frac{1}{4}\right)^2$, or $\frac{1}{16}$.

Describe the effect of each change on the given measurement of the figure.

7. surface area
The radius is multiplied by 4.

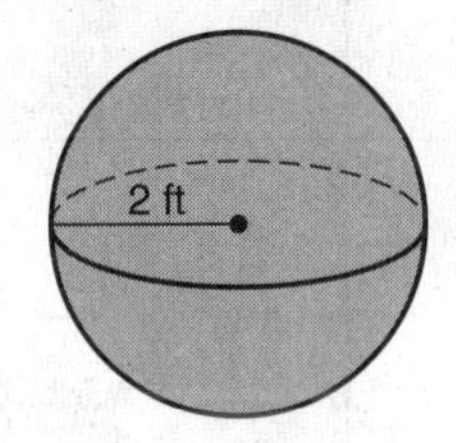

____

8. volume
The dimensions are multiplied by $\frac{1}{2}$.

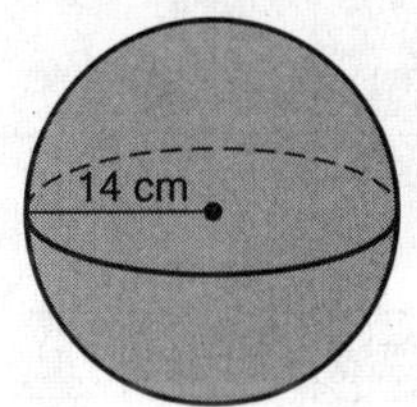

____

Find the surface area and volume of each composite figure. Round to the nearest tenth.

9. *Hint:* To find the surface area, add the lateral area of the cylinder, the area of one base, and the surface area of the hemisphere.

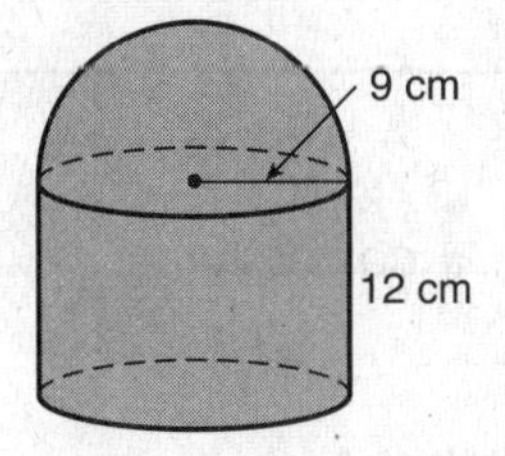

____

10. *Hint:* To find the volume, subtract the volume of the hemisphere from the volume of the cylinder.

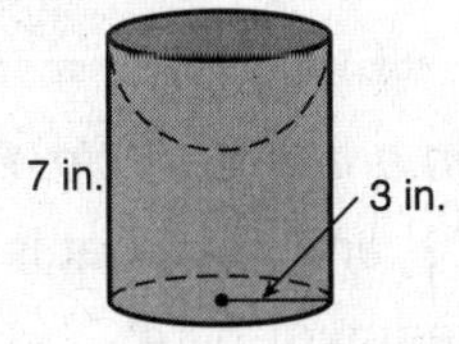

____

Copyright © by Holt, Rinehart and Winston.
All rights reserved.

Name ______________________ Date ____________ Class ____________

LESSON 10-8

Challenge

Spheres, Cylinders, and Archimedes

The Greek mathematician Archimedes (ca. 287–212 B.C.), a native of Syracuse, Sicily, is considered one of the greatest mathematicians of all time. He is perhaps best known for his contributions to the field of mechanics, such as the invention of the Archimedean screw and the discovery of the principle of buoyancy. However, it was geometry that Archimedes found most fascinating, and he established the first exact expressions for the volume and surface area of a sphere.

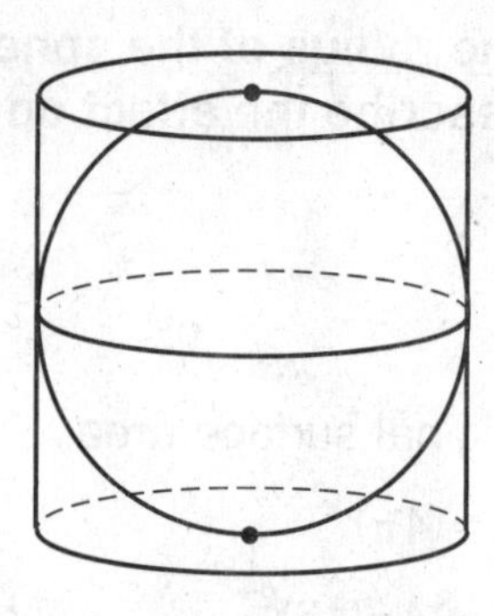

When a sphere is inscribed in a right cylinder, a diameter of the sphere lies on the lateral surface of the cylinder, and the sphere intersects the cylinder at the centers of the base.

In a work titled *On the Sphere and Cylinder,* Archimedes examined the relationship between a right cylinder and a sphere inscribed in it. He considered this work to be so significant that he requested a representation of a cylinder and inscribed sphere to be engraved on his tombstone.

Refer to the figure above. Find each measure for the given radius. Give answers in exact form.

	Radius	Lateral Area of Cylinder	Surface Area of Cylinder	Volume of Cylinder	Surface Area of Sphere	Volume of Sphere
1.	2 in.					
2.	5 cm					
3.	10 ft					
4.	1.5 m					

5. Refer to the table in Exercises 1–4. Archimedes discovered that two of the sets of measures related the sphere to the cylinder by the ratio 2 : 3. Which measures are they?

____

6. Refer to your answer to Exercise 5. On a separate sheet of paper, demonstrate algebraically why the two ratios you identified are always 2 : 3.

____

7. In his work *Measurement of a Circle,* Archimedes demonstrated that the value of π was between $\frac{223}{71}$ and $\frac{22}{7}$. That is, he approximated the value of π correctly to the nearest hundredth, or 3.14.

   Using your library or the Internet as a resource, research the method that Archimedes used to make his approximation. Write your results on a separate sheet of paper.

Copyright © by Holt, Rinehart and Winston. All rights reserved.

Name ______________________ Date ____________ Class ____________

LESSON 10-8

Problem Solving
Spheres

1. A globe has a volume of 288π in^3. What is the surface area of the globe? Give your answer in terms of π.

____

2. Eight bocce balls are in a box 18 inches long, 9 inches wide, and 4.5 inches deep. If each ball has a diameter of 4.5 inches, what is the volume of the space around the balls? Round to the nearest tenth.

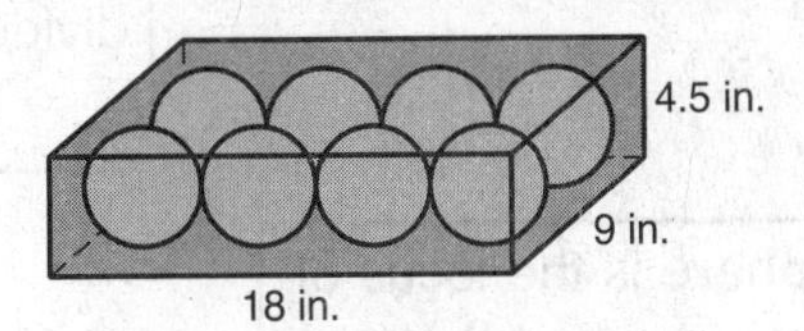

____

Use the table for Exercises 3 and 4.

Ganymede, one of Jupiter's moons, is the largest moon in the solar system.

Moon	Diameter
Earth's moon	2160 mi
Ganymede	3280 mi

3. Approximately how many times as great as the volume of Earth's moon is the volume of Ganymede?

____

4. Approximately how many times as great is the surface area of Ganymede than the surface area of Earth's moon?

____

Choose the best answer.

5. What is the volume of a sphere with a great circle that has an area of 225π cm^2?

A 300π cm^3
B 900π cm^3
C 2500π cm^3
D 4500π cm^3

6. A hemisphere has a surface area of 972π cm^2. If the radius is multiplied by $\frac{1}{3}$, what will be the surface area of the new hemisphere?

F 36π cm^2
G 108π cm^2
H 162π cm^2
J 324π cm^2

7. Which expression represents the volume of the composite figure formed by the hemisphere and cone?

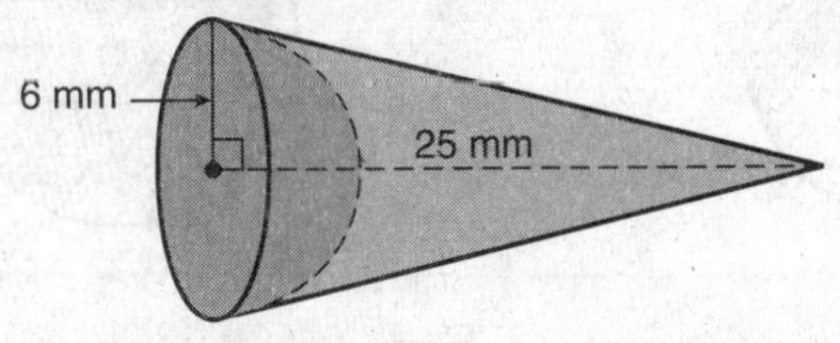

A 52π mm^3
B 156π mm^3
C 276π mm^3
D 288π mm^3

8. Which best represents the surface area of the composite figure?

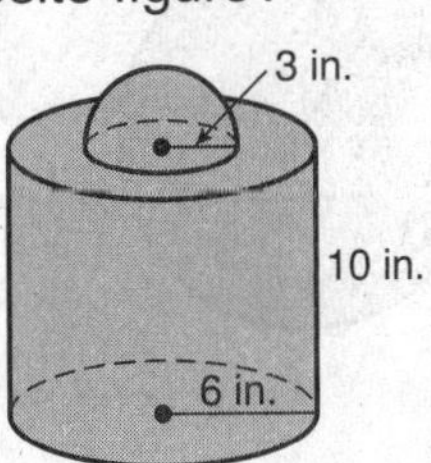

F 129π in^2
G 138π in^2
H 201π in^2
J 210π in^2

Copyright © by Holt, Rinehart and Winston. All rights reserved.

Name ______________________ Date __________ Class __________

LESSON 10-8

Reading Strategies

Focus on Vocabulary

The diagram below describes the parts of a sphere and gives you the formulas for surface area and volume of a sphere.

A **hemisphere** is half a sphere. A **great circle** divides a sphere into two hemispheres.

A **sphere** is the locus of points in space that are a fixed distance from the **center** of the sphere.

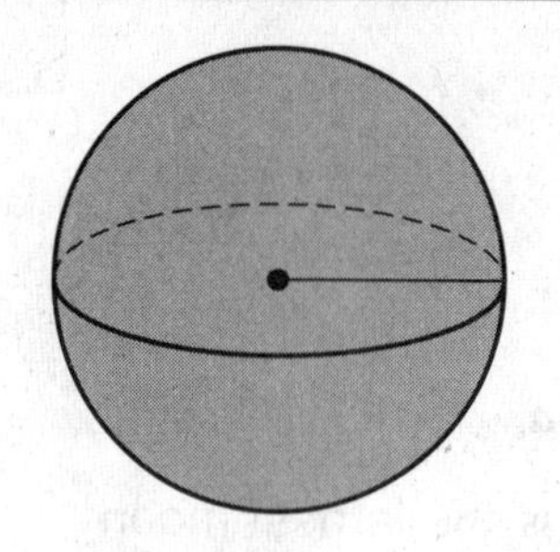

A **radius** r connects the center of the sphere to any point on the sphere.

The formula for the **volume** of a sphere is $V = \frac{4}{3}\pi r^3$.

The formula for the **surface area** of a sphere is $S = 4\pi r^2$.

Answer the following.

1. The ______________ of a sphere connects the center of the sphere to any point on the sphere.
2. A(n) ______________ is half a sphere.
3. The radius of a sphere is 6 centimters. What is the radius of its great circle? ______________
4. The volume of a sphere is 900π ft^3. What is the volume of one of its hemispheres? ______________

Find the surface area and volume of each sphere. Give your answers in terms of π.

5. 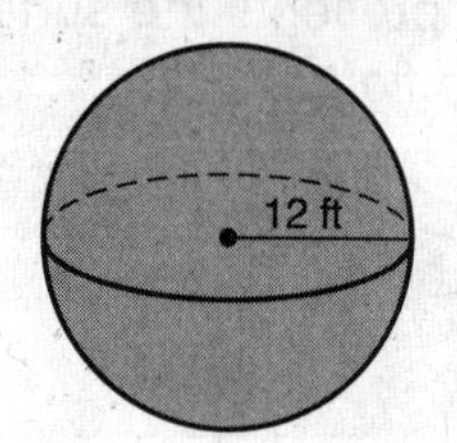

$S =$ ______________

$V =$ ______________

6. 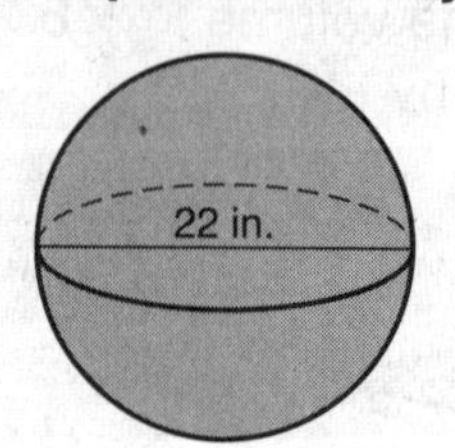

$S =$ ______________

$V =$ ______________

7. 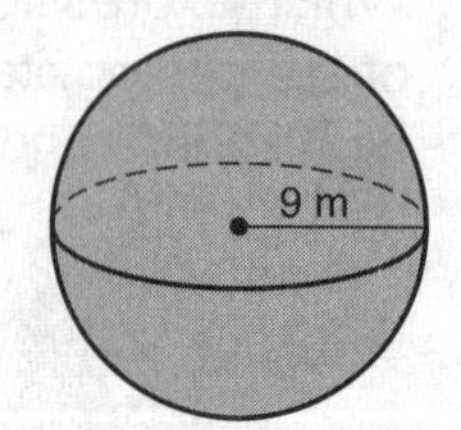

$S =$ ______________

$V =$ ______________

Copyright © by Holt, Rinehart and Winston.
All rights reserved.

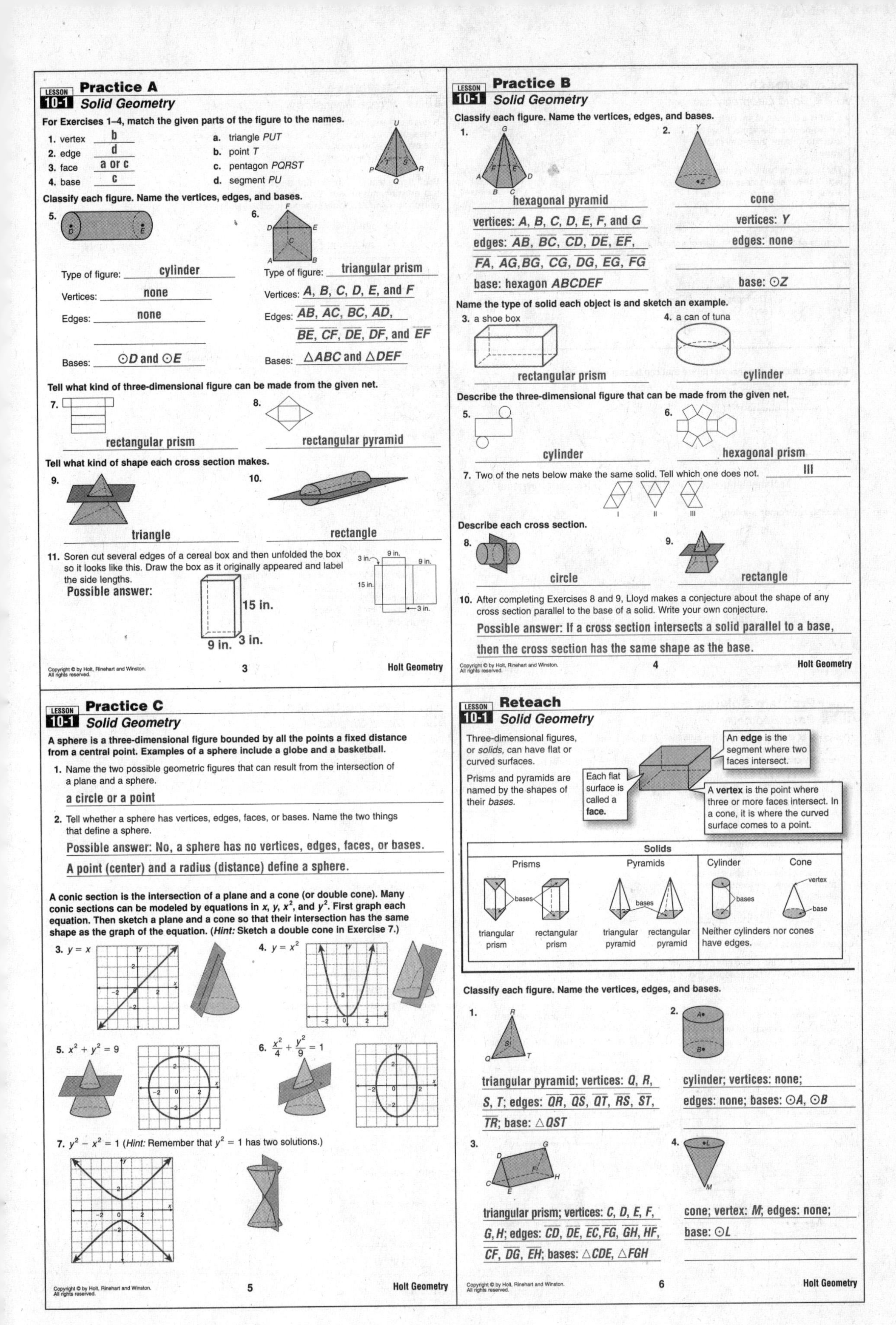

LESSON 10-1

Practice A

Solid Geometry

For Exercises 1–4, match the given parts of the figure to the names.

1. vertex b
2. edge d
3. face a or c
4. base c

a. triangle *PUT*
b. point *T*
c. pentagon *PQRST*
d. segment *PU*

Classify each figure. Name the vertices, edges, and bases.

5. Type of figure: cylinder
Vertices: none
Edges: none
Bases: ⊙*D* and ⊙*E*

6. Type of figure: triangular prism
Vertices: *A, B, C, D, E,* and *F*
Edges: $\overline{AB}, \overline{AC}, \overline{BC}, \overline{AD}, \overline{BE}, \overline{CF}, \overline{DE}, \overline{DF}$, and $\overline{EF}$
Bases: △*ABC* and △*DEF*

Tell what kind of three-dimensional figure can be made from the given net.

7. rectangular prism
8. rectangular pyramid

Tell what kind of shape each cross section makes.

9. triangle
10. rectangle

11. Soren cut several edges of a cereal box and then unfolded the box so it looks like this. Draw the box as it originally appeared and label the side lengths.
Possible answer: 15 in., 9 in., 3 in.

Copyright © by Holt, Rinehart and Winston. All rights reserved. 3 Holt Geometry

LESSON 10-1

Practice B

Solid Geometry

Classify each figure. Name the vertices, edges, and bases.

1. hexagonal pyramid
vertices: *A, B, C, D, E, F,* and *G*
edges: $\overline{AB}, \overline{BC}, \overline{CD}, \overline{DE}, \overline{EF}, \overline{FA}, \overline{AG}, \overline{BG}, \overline{CG}, \overline{DG}, \overline{EG}, \overline{FG}$
base: hexagon *ABCDEF*

2. cone
vertices: *Y*
edges: none
base: ⊙*Z*

Name the type of solid each object is and sketch an example.

3. a shoe box — rectangular prism
4. a can of tuna — cylinder

Describe the three-dimensional figure that can be made from the given net.

5. cylinder
6. hexagonal prism

7. Two of the nets below make the same solid. Tell which one does not. III

Describe each cross section.

8. circle
9. rectangle

10. After completing Exercises 8 and 9, Lloyd makes a conjecture about the shape of any cross section parallel to the base of a solid. Write your own conjecture.
Possible answer: If a cross section intersects a solid parallel to a base, then the cross section has the same shape as the base.

Copyright © by Holt, Rinehart and Winston. All rights reserved. 4 Holt Geometry

LESSON 10-1

Practice C

Solid Geometry

A sphere is a three-dimensional figure bounded by all the points a fixed distance from a central point. Examples of a sphere include a globe and a basketball.

1. Name the two possible geometric figures that can result from the intersection of a plane and a sphere.
a circle or a point

2. Tell whether a sphere has vertices, edges, faces, or bases. Name the two things that define a sphere.
Possible answer: No, a sphere has no vertices, edges, faces, or bases. A point (center) and a radius (distance) define a sphere.

A conic section is the intersection of a plane and a cone (or double cone). Many conic sections can be modeled by equations in *x*, *y*, x^2, and y^2. First graph each equation. Then sketch a plane and a cone so that their intersection has the same shape as the graph of the equation. (*Hint:* Sketch a double cone in Exercise 7.)

3. $y = x$
4. $y = x^2$
5. $x^2 + y^2 = 9$
6. $\frac{x^2}{4} + \frac{y^2}{9} = 1$
7. $y^2 - x^2 = 1$ (*Hint:* Remember that $y^2 = 1$ has two solutions.)

Copyright © by Holt, Rinehart and Winston. All rights reserved. 5 Holt Geometry

LESSON 10-1

Reteach

Solid Geometry

Three-dimensional figures, or *solids,* can have flat or curved surfaces.

Prisms and pyramids are named by the shapes of their *bases.*

Each flat surface is called a **face.**

An **edge** is the segment where two faces intersect.

A **vertex** is the point where three or more faces intersect. In a cone, it is where the curved surface comes to a point.

Solids					
Prisms		Pyramids		Cylinder	Cone
triangular prism	rectangular prism	triangular pyramid	rectangular pyramid	Neither cylinders nor cones have edges.	

Classify each figure. Name the vertices, edges, and bases.

1. triangular pyramid; vertices: *Q, R, S, T*; edges: $\overline{QR}, \overline{QS}, \overline{QT}, \overline{RS}, \overline{ST}, \overline{TR}$; base: △*QST*

2. cylinder; vertices: none; edges: none; bases: ⊙*A*, ⊙*B*

3. triangular prism; vertices: *C, D, E, F, G, H*; edges: $\overline{CD}, \overline{DE}, \overline{EC}, \overline{FG}, \overline{GH}, \overline{HF}, \overline{CF}, \overline{DG}, \overline{EH}$; bases: △*CDE*, △*FGH*

4. cone; vertex: *M*; edges: none; base: ⊙*L*

Copyright © by Holt, Rinehart and Winston. All rights reserved. 6 Holt Geometry

Copyright © by Holt, Rinehart and Winston.
All rights reserved.

LESSON 10-1 Reteach
Solid Geometry continued

A **net** is a diagram of the surfaces of a three-dimensional figure. It can be folded to form the three-dimensional figure.

The net at right has one rectangular face. The remaining faces are triangles, and so the net forms a rectangular pyramid.

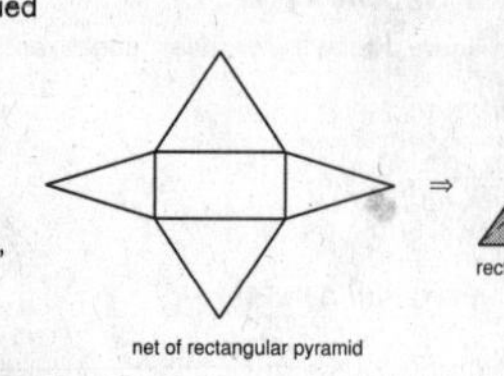

A **cross section** is the intersection of a three-dimensional figure and a plane.

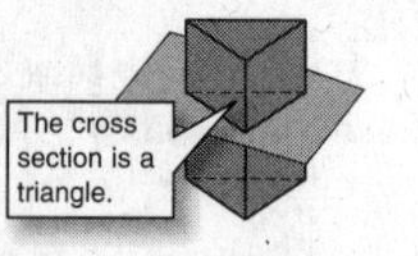

Describe the three-dimensional figure that can be made from the given net.

5. 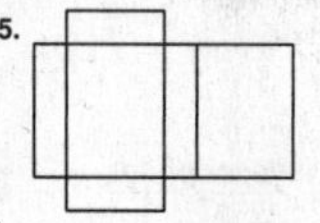

rectangular prism

6. 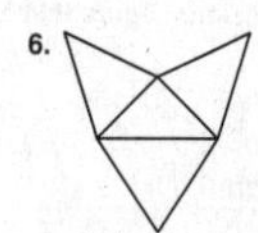

triangular pyramid

Describe each cross section.

7. 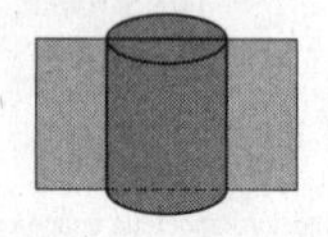

rectangle

8. 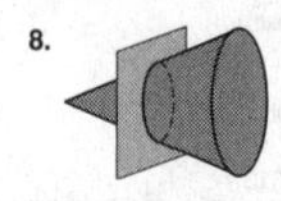

circle

Copyright © by Holt, Rinehart and Winston. All rights reserved.

LESSON 10-1 Challenge
Three-Dimensional Pentominoes

A **three-dimensional pentomino** is a figure formed by five identical cubes arranged so that each cube shares a common face with at least one other cube. Knowledge of three-dimensional symmetries can be helpful in working with these pentominoes.

Each figure is a three-dimensional pentomino. Tell whether it has *reflectional symmetry only, rotational symmetry only, both reflectional and rotational symmetry,* or *no symmetry.*

1. 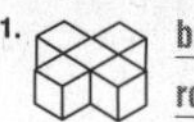 both reflectional and rotational symmetry

2. 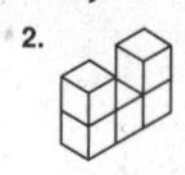 reflectional symmetry only

3.  no symmetry

4. 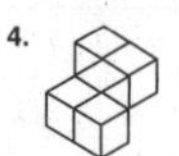 rotational symmetry only

In Exercises 5–7 a three-dimensional pentomino is shown at left. Name the pentomino A, B, or C to its right that is identical to it.

5. 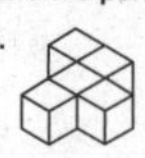 A. 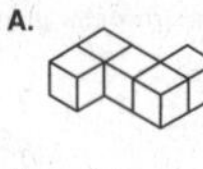 B. 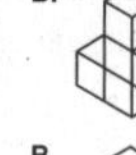 C. 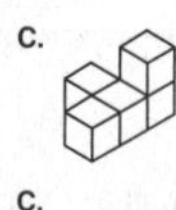 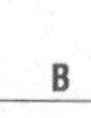 B

6. 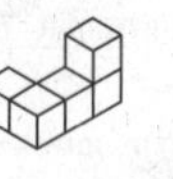 A. 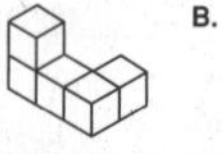 B. 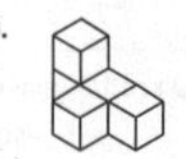 C. 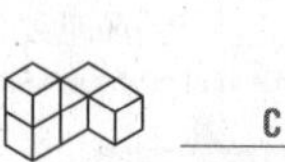 C

7. A.  B. 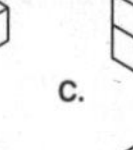 C. 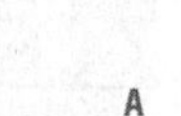 A

8. Create an original puzzle: Find a way to form a right rectangular prism by combining two or more three-dimensional pentominoes. Then make models of the pentominoes. Trade these "puzzle pieces" with a partner and see who can solve the other's puzzle first.

Sample answer:

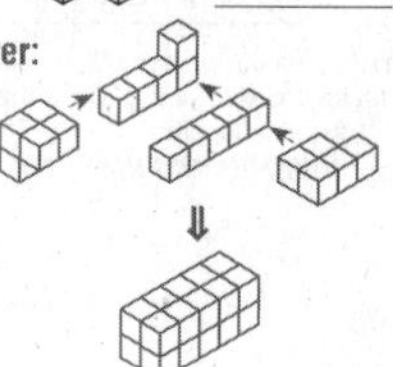

Puzzles will vary.

Copyright © by Holt, Rinehart and Winston. All rights reserved.

LESSON 10-1 Problem Solving
Solid Geometry

1. A slice of cheese is cut from the cylinder-shaped cheese as shown. Describe the cross section.

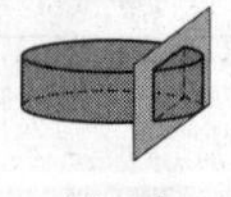

rectangle

2. Mara has cut out five pieces of fabric to sew together to form a pillow. There are three rectangular pieces and two triangles. Describe the solid that will be formed.

triangular prism

3. A square pyramid is intersected by a plane as shown. Describe the cross section.

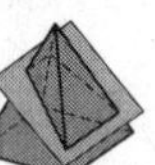

trapezoid

Choose the best answer.

4. A gift box is in the shape of a pentagonal prism. How many faces, edges, and vertices does the box have?
   - A 6 faces, 10 edges, 6 vertices
   - B 7 faces, 12 edges, 10 vertices
   - (C) 7 faces, 15 edges, 10 vertices
   - D 8 faces, 18 edges, 12 vertices

5. Which two solids have the same number of vertices?
   - F rectangular prism and triangular pyramid
   - G triangular prism and rectangular pyramid
   - H rectangular prism and pentagonal pyramid
   - (J) triangular prism and pentagonal pyramid

6. Which three-dimensional figure does the net represent?
   - (A)
   - B
   - C
   - D

7. Which can be a true statement about the triangular prism whose net is shown?

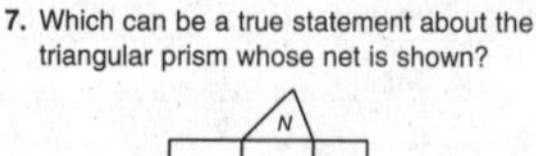
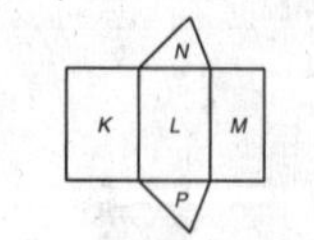

   - F Faces L and M are perpendicular.
   - G Faces N and P are perpendicular.
   - H Faces K and L are parallel.
   - (J) Faces N and P are parallel.

Copyright © by Holt, Rinehart and Winston. All rights reserved.

LESSON 10-1 Reading Strategies
Use a Graphic Aid

Solids are made up of flat surfaces and curved surfaces. A flat surface is called a **face.** The intersection of two faces is a segment called an **edge.** Three or more faces intersect at a **vertex.** Use the graphic aid below to understand solids.

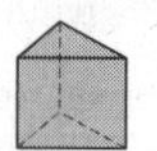

- Two parallel congruent polygonal bases
- The bases are connected by faces that are parallelograms.
- Named for the shape of its base

- One polygonal base
- The triangular faces meet at a common vertex.
- Named for the shape of its base

Prism	Pyramid
Solids	
Cylinder	Cone

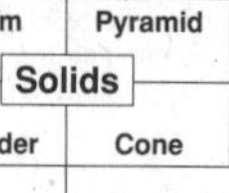

- Two parallel congruent circular bases
- A curved surface connects the bases.

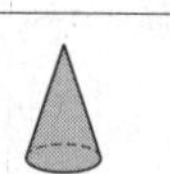

- One circular base
- A curved surface connects the base to a vertex.

Answer the following.

1. Name two solids that have circular bases.

   cylinder   cone

2. A(n) prism has two parallel congruent polygonal bases.

3. The intersection of two faces is called a(n) edge.

Identify each solid.

4. 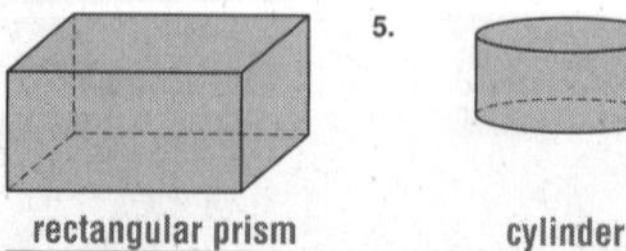

rectangular prism

5. 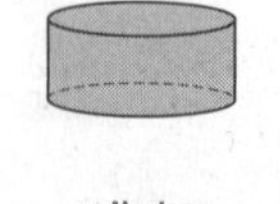

cylinder

6. triangular pyramid

Copyright © by Holt, Rinehart and Winston. All rights reserved.

Copyright © by Holt, Rinehart and Winston. All rights reserved.

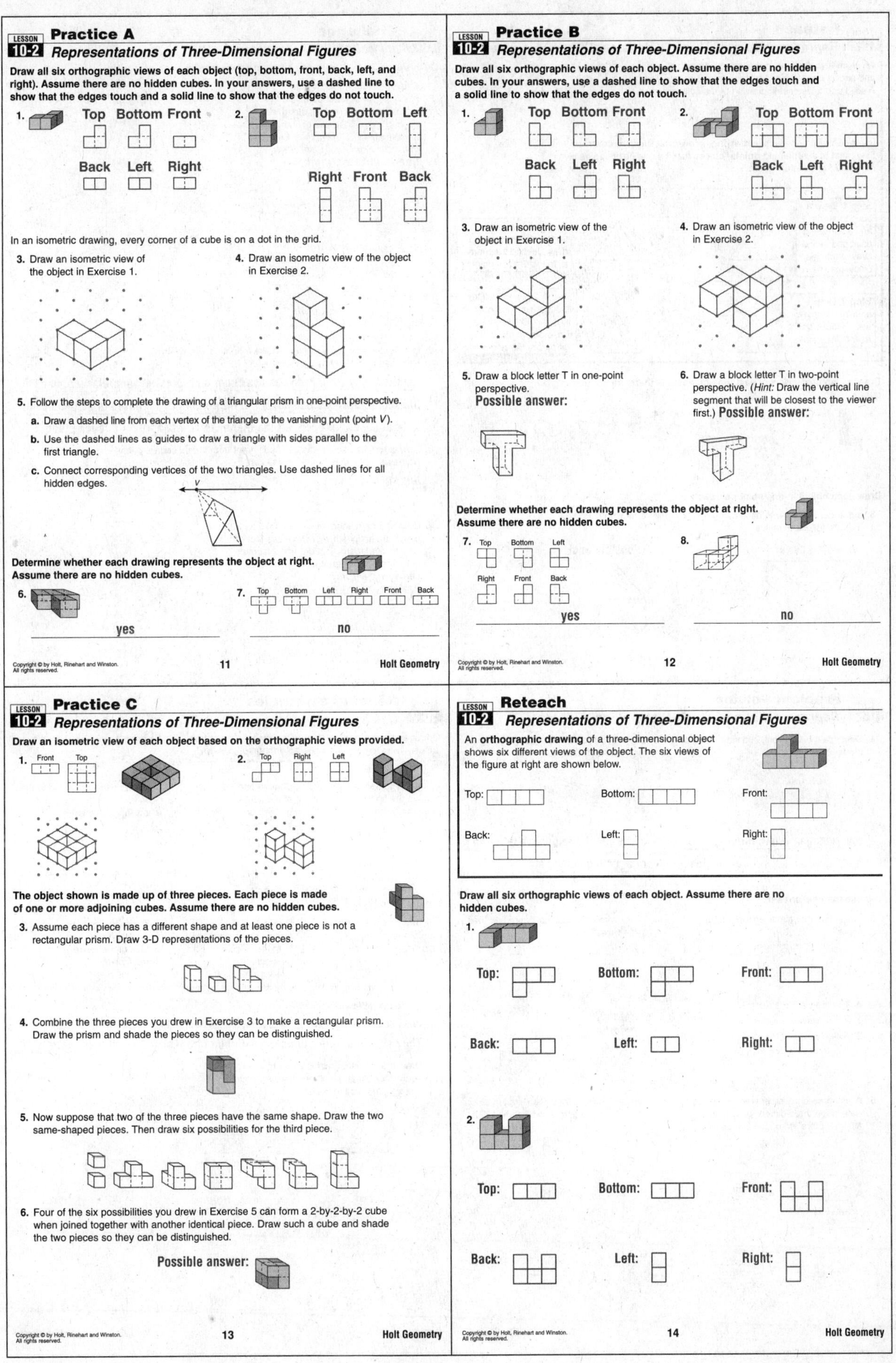

LESSON 10-2

Practice A

Representations of Three-Dimensional Figures

Draw all six orthographic views of each object (top, bottom, front, back, left, and right). Assume there are no hidden cubes. In your answers, use a dashed line to show that the edges touch and a solid line to show that the edges do not touch.

1. Top Bottom Front
Back Left Right

2. Top Bottom Left
Right Front Back

In an isometric drawing, every corner of a cube is on a dot in the grid.

3. Draw an isometric view of the object in Exercise 1.

4. Draw an isometric view of the object in Exercise 2.

5. Follow the steps to complete the drawing of a triangular prism in one-point perspective.
   a. Draw a dashed line from each vertex of the triangle to the vanishing point (point *V*).
   b. Use the dashed lines as guides to draw a triangle with sides parallel to the first triangle.
   c. Connect corresponding vertices of the two triangles. Use dashed lines for all hidden edges.

V

Determine whether each drawing represents the object at right. Assume there are no hidden cubes.

6. yes

7. Top Bottom Left Right Front Back
no

Copyright © by Holt, Rinehart and Winston. All rights reserved.

LESSON 10-2

Practice B

Representations of Three-Dimensional Figures

Draw all six orthographic views of each object. Assume there are no hidden cubes. In your answers, use a dashed line to show that the edges touch and a solid line to show that the edges do not touch.

1. Top Bottom Front
Back Left Right

2. Top Bottom Front
Back Left Right

3. Draw an isometric view of the object in Exercise 1.

4. Draw an isometric view of the object in Exercise 2.

5. Draw a block letter T in one-point perspective.
Possible answer:

6. Draw a block letter T in two-point perspective. (*Hint:* Draw the vertical line segment that will be closest to the viewer first.) **Possible answer:**

Determine whether each drawing represents the object at right. Assume there are no hidden cubes.

7. Top Bottom Left
Right Front Back
yes

8. no

Copyright © by Holt, Rinehart and Winston. All rights reserved.

LESSON 10-2

Practice C

Representations of Three-Dimensional Figures

Draw an isometric view of each object based on the orthographic views provided.

1. Front Top

2. Top Right Left

The object shown is made up of three pieces. Each piece is made of one or more adjoining cubes. Assume there are no hidden cubes.

3. Assume each piece has a different shape and at least one piece is not a rectangular prism. Draw 3-D representations of the pieces.

4. Combine the three pieces you drew in Exercise 3 to make a rectangular prism. Draw the prism and shade the pieces so they can be distinguished.

5. Now suppose that two of the three pieces have the same shape. Draw the two same-shaped pieces. Then draw six possibilities for the third piece.

6. Four of the six possibilities you drew in Exercise 5 can form a 2-by-2-by-2 cube when joined together with another identical piece. Draw such a cube and shade the two pieces so they can be distinguished.

Possible answer:

Copyright © by Holt, Rinehart and Winston. All rights reserved.

LESSON 10-2

Reteach

Representations of Three-Dimensional Figures

An **orthographic drawing** of a three-dimensional object shows six different views of the object. The six views of the figure at right are shown below.

Top: Bottom: Front:
Back: Left: Right:

Draw all six orthographic views of each object. Assume there are no hidden cubes.

1. Top: Bottom: Front:
Back: Left: Right:

2. Top: Bottom: Front:
Back: Left: Right:

Copyright © by Holt, Rinehart and Winston. All rights reserved.

Copyright © by Holt, Rinehart and Winston.
All rights reserved.

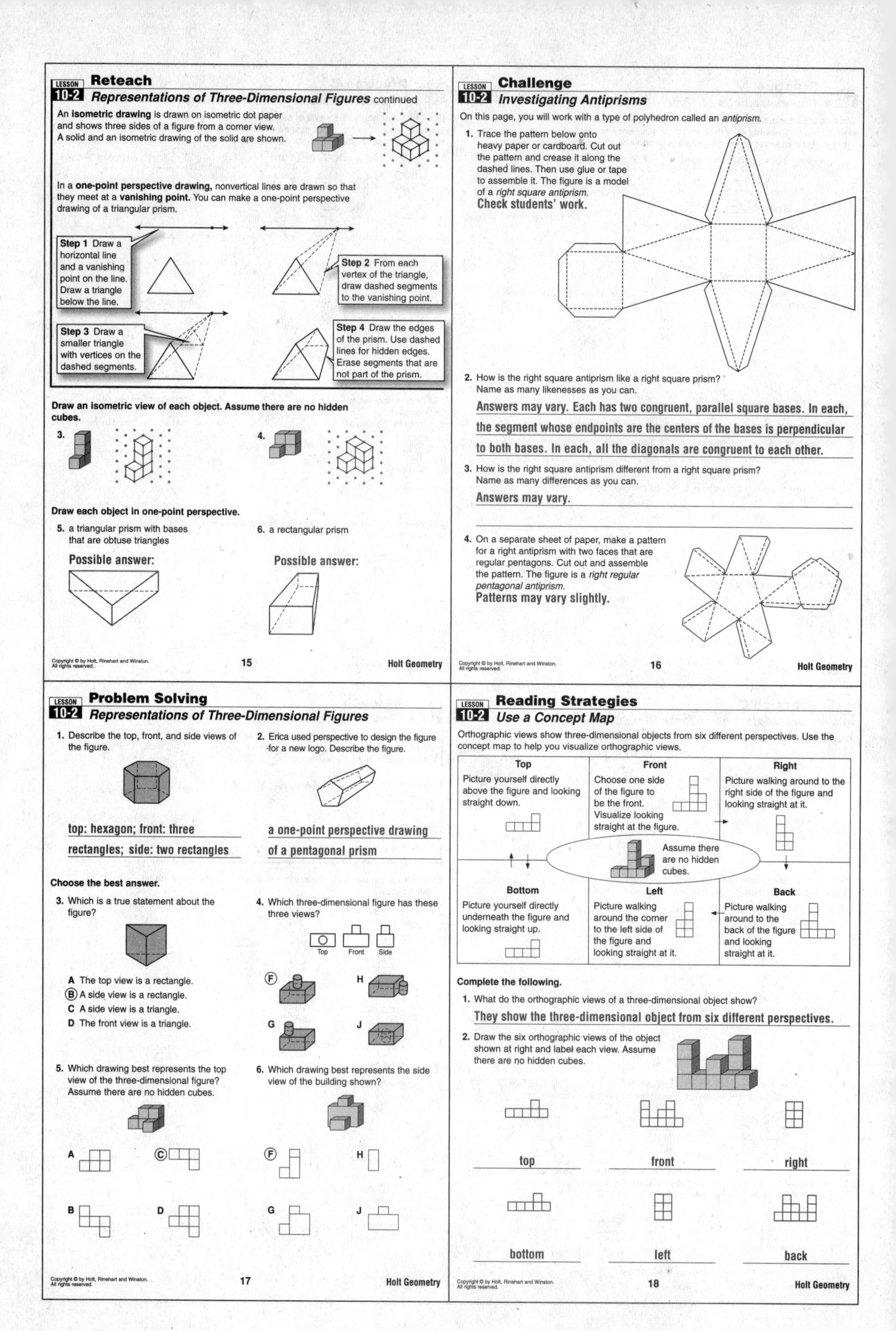

LESSON 10-2

Reteach

Representations of Three-Dimensional Figures continued

An **isometric drawing** is drawn on isometric dot paper and shows three sides of a figure from a corner view. A solid and an isometric drawing of the solid are shown.

In a **one-point perspective drawing,** nonvertical lines are drawn so that they meet at a **vanishing point.** You can make a one-point perspective drawing of a triangular prism.

Step 1 Draw a horizontal line and a vanishing point on the line. Draw a triangle below the line.

Step 2 From each vertex of the triangle, draw dashed segments to the vanishing point.

Step 3 Draw a smaller triangle with vertices on the dashed segments.

Step 4 Draw the edges of the prism. Use dashed lines for hidden edges. Erase segments that are not part of the prism.

Draw an isometric view of each object. Assume there are no hidden cubes.

3.

4.

Draw each object in one-point perspective.

5. a triangular prism with bases that are obtuse triangles

Possible answer:

6. a rectangular prism

Possible answer:

Copyright © by Holt, Rinehart and Winston. All rights reserved.

LESSON 10-2

Challenge

Investigating Antiprisms

On this page, you will work with a type of polyhedron called an *antiprism.*

1. Trace the pattern below onto heavy paper or cardboard. Cut out the pattern and crease it along the dashed lines. Then use glue or tape to assemble it. The figure is a model of a *right square antiprism.*

Check students' work.

2. How is the right square antiprism like a right square prism? Name as many likenesses as you can.

Answers may vary. Each has two congruent, parallel square bases. In each, the segment whose endpoints are the centers of the bases is perpendicular to both bases. In each, all the diagonals are congruent to each other.

3. How is the right square antiprism different from a right square prism? Name as many differences as you can.

Answers may vary.

4. On a separate sheet of paper, make a pattern for a right antiprism with two faces that are regular pentagons. Cut out and assemble the pattern. The figure is a *right regular pentagonal antiprism.*

Patterns may vary slightly.

Copyright © by Holt, Rinehart and Winston. All rights reserved.

LESSON 10-2

Problem Solving

Representations of Three-Dimensional Figures

1. Describe the top, front, and side views of the figure.

top: hexagon; front: three rectangles; side: two rectangles

2. Erica used perspective to design the figure for a new logo. Describe the figure.

a one-point perspective drawing of a pentagonal prism

Choose the best answer.

3. Which is a true statement about the figure?

A The top view is a rectangle.
(B) A side view is a rectangle.
C A side view is a triangle.
D The front view is a triangle.

4. Which three-dimensional figure has these three views?

Top Front Side

(F) H G J

5. Which drawing best represents the top view of the three-dimensional figure? Assume there are no hidden cubes.

A (C) B D

6. Which drawing best represents the side view of the building shown?

(F) H G J

Copyright © by Holt, Rinehart and Winston. All rights reserved.

LESSON 10-2

Reading Strategies

Use a Concept Map

Orthographic views show three-dimensional objects from six different perspectives. Use the concept map to help you visualize orthographic views.

Top
Picture yourself directly above the figure and looking straight down.

Front
Choose one side of the figure to be the front. Visualize looking straight at the figure.

Right
Picture walking around to the right side of the figure and looking straight at it.

Assume there are no hidden cubes.

Bottom
Picture yourself directly underneath the figure and looking straight up.

Left
Picture walking around the corner to the left side of the figure and looking straight at it.

Back
Picture walking around to the back of the figure and looking straight at it.

Complete the following.

1. What do the orthographic views of a three-dimensional object show?

They show the three-dimensional object from six different perspectives.

2. Draw the six orthographic views of the object shown at right and label each view. Assume there are no hidden cubes.

top **front** **right**

bottom **left** **back**

Copyright © by Holt, Rinehart and Winston. All rights reserved.

Copyright © by Holt, Rinehart and Winston. All rights reserved.

LESSON 10-3
Practice A
Formulas in Three Dimensions

Match the letter of each formula to its name.

1. Euler's Formula ___b___
2. diagonal of a rectangular prism ___c___
3. distance in three dimensions ___d___
4. midpoint in three dimensions ___a___

a. $M\left(\frac{x_1+x_2}{2}, \frac{y_1+y_2}{2}, \frac{z_1+z_2}{2}\right)$

b. $V - E + F = 2$

c. $d = \sqrt{\ell^2 + w^2 + h^2}$

d. $d = \sqrt{(x_2-x_1)^2 + (y_2-y_1)^2 + (z_2-z_1)^2}$

Count the number of vertices, edges, and faces of each polyhedron. Use your results to verify Euler's Formula.

5. 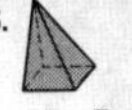

$V = 5; E = 8; F = 5; 5 - 8 + 5 = 2$

6. 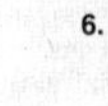

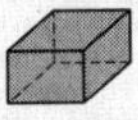

$V = 8; E = 12; F = 6; 8 - 12 + 6 = 2$

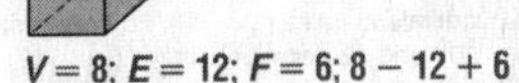

For Exercises 7–9, use the formula for the length of a diagonal to find the unknown dimension in each polyhedron. Round to the nearest tenth.

7. the length of a diagonal of a cube with edge length 3 in. ___5.2 in.___
8. the length of a diagonal of a 7-cm-by-10-cm-by-4-cm rectangular prism ___12.8 cm___
9. the height of a rectangular prism with a 6-m-by-6-m base and a 9 m diagonal ___3 m___
10. A rectangular prism with length 3, width 2, and height 4 has one vertex at (0, 0, 0). Three other vertices are at (3, 0, 0), (0, 2, 0), and (0, 0, 4). Find the four other vertices. Then graph the figure.

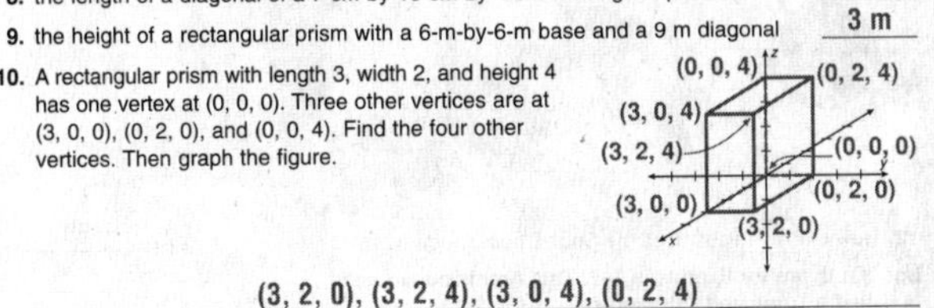

(3, 2, 0), (3, 2, 4), (3, 0, 4), (0, 2, 4)

Use the formula for distance in three dimensions to find the distance between the given points. Use the midpoint formula in three dimensions to find the midpoint of the segment with the given endpoints. Round to the nearest tenth if necessary.

11. (0, 0, 0) and (2, 4, 6)

7.5 units; (1, 2, 3)

12. (1, 0, 5) and (0, 4, 0)

6.5 units; (0.5, 2, 2.5)

13. The world's largest ball of twine wound by a single individual weighs 17,400 pounds and has a 12-foot diameter. Roman climbs on top of the ball for a picture. To take the best picture, Lysandra moves 15 feet back and then 6 feet to her right. Find the distance from Lysandra to Roman. Round to the nearest tenth. ___24.9 feet___

Copyright © by Holt, Rinehart and Winston. All rights reserved.

LESSON 10-3
Practice B
Formulas in Three Dimensions

Find the number of vertices, edges, and faces of each polyhedron. Use your results to verify Euler's Formula.

1. 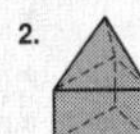

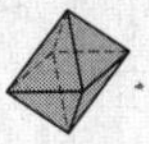

$V = 6; E = 12; F = 8;$
$6 - 12 + 8 = 2$

2.

$V = 7; E = 12; F = 7;$
$7 - 12 + 7 = 2$

Find the unknown dimension in each polyhedron. Round to the nearest tenth.

3. the edge length of a cube with a diagonal of 9 ft ___5.2 ft___
4. the length of a diagonal of a 15-mm-by-20-mm-by-8-mm rectangular prism ___26.2 mm___
5. the length of a rectangular prism with width 2 in., height 18 in., and a 21-in. diagonal ___10.6 in.___

Graph each figure.

6. a square prism with base edge length 4 units, height 2 units, and one vertex at (0, 0, 0)

Possible answer:

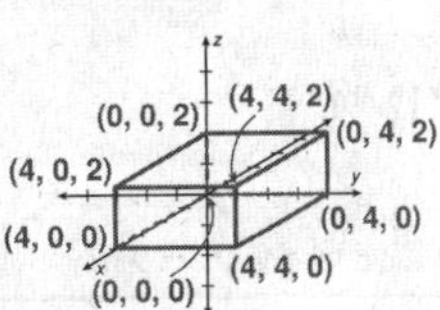
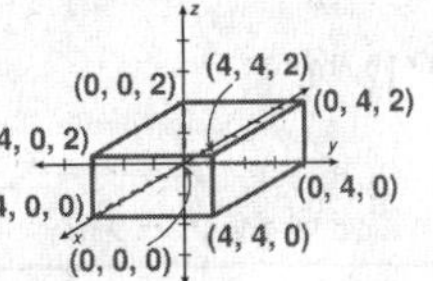

7. a cone with base diameter 6 units, height 3 units, and base centered at (0, 0, 0)

Possible answer:

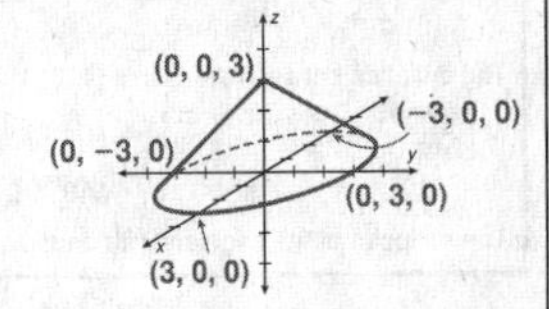

Find the distance between the given points. Find the midpoint of the segment with the given endpoints. Round to the nearest tenth if necessary.

8. (1, 10, 3) and (5, 5, 5)

6.7 units; (3, 7.5, 4)

9. (−8, 0, 11) and (2, −6, −17)

30.3 units; (−3, −3, −3)

Copyright © by Holt, Rinehart and Winston. All rights reserved.

LESSON 10-3
Practice C
Formulas in Three Dimensions

1. The distance from (0, 0, 0) to the surface of a solid is 4 units. Graph the solid.

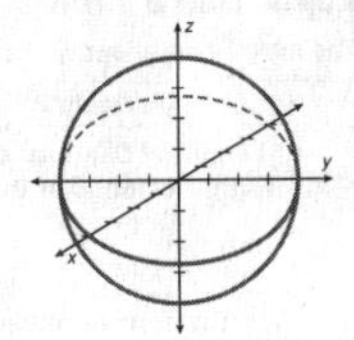

2. Each edge of the solid shown in the figure measures 5 in. Find the length of $\overline{AB}$. Give an exact answer and an answer rounded to the nearest tenth.

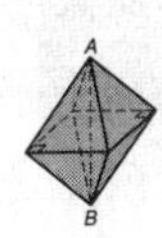

$5\sqrt{2}$ in.; 7.1 in.

3. Find the length of $\overline{AB}$ if the bipyramid in Exercise 2 were based on a triangle rather than on a square. Round to the nearest tenth. ___8.2 in.___
4. Find the length of $\overline{AB}$ if the bipyramid in Exercise 2 were based on a pentagon rather than on a square. Round to the nearest tenth. ___5.3 in.___
5. If the bipyramid in Exercise 2 were based on a hexagon instead of a square, describe what sort of shape would result. Explain your answer.

The shape would be a flat hexagon; possible answer: The distance to the vertex of the bipyramid from the midpoint of a side (the slant height) would be $\frac{5\sqrt{3}}{2}$ in. The distance from the midpoint of a side to the center of the hexagon (the apothem) would also be $\frac{5\sqrt{3}}{2}$ in. Therefore, the height *AB* would be zero.

6. The distance from *A*(−2, 7, 0) to *B*(3, 2, *b*) and from *A* to *C*(3, 2, *c*) is 10 units. *D* lies on $\overline{BC}$ so that *AD* is the shortest distance from *A* to $\overline{BC}$. Find the coordinates of *D* without calculating. Explain how you got the answer.

D(3, 2, 0); possible answer: Because *B* and *C* have the same *x*- and *y*-coordinates, *D* must also have those *x*- and *y*-coordinates to lie on $\overline{BC}$. Any difference in length from *A* to $\overline{BC}$ is caused by changes in the *z*-coordinate, and the shortest distance occurs when *D* has the same *z*-coordinate as *A*.

7. A rectangular prism has vertices, in no particular order, at (−10, 8, 2), (−15, 8, 10), (−10, 5, 10), (−10, 5, 2), (−10, 8, 10), (−15, 5, 2), (−15, 5, 10), and (−15, 8, 2). Find the length of a diagonal of the prism. Round to the nearest tenth. ___$7\sqrt{2} \approx 9.9$ units___
8. Find the coordinates of a point that is equidistant from each of the eight vertices of the prism in Exercise 7. ___(−12.5, 6.5, 6)___

Tyrone has eight 1-in. cubes. He arranges all eight of them to make different rectangular prisms. Find the dimensions of the prisms based on the diagonal lengths given below.

9. $\sqrt{66}$ in.

an 8-by-1-by-1 prism

10. $\sqrt{21}$ in.

a 4-by-2-by-1 prism

11. $2\sqrt{3}$ in.

a 2-by-2-by-2 prism

Copyright © by Holt, Rinehart and Winston. All rights reserved.

LESSON 10-3
Reteach
Formulas in Three Dimensions

A **polyhedron** is a solid formed by four or more polygons that intersect only at their edges. Prisms and pyramids are polyhedrons. Cylinders and cones are not.

Euler's Formula

For any polyhedron with *V* vertices, *E* edges, and *F* faces,

$V - E + F = 2.$

Example

$V - E + F = 2$ Euler's Formula

$4 - 6 + 4 = 2$ $V = 4, E = 6, F = 4$

$2 = 2$

4 vertices, 6 edges, 4 faces

Diagonal of a Right Rectangular Prism

The length of a diagonal *d* of a right rectangular prism with length ℓ, width *w*, and height *h* is

$d = \sqrt{\ell^2 + w^2 + h^2}.$

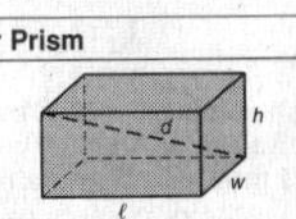

Find the height of a rectangular prism with a 4 cm by 3 cm base and a 7 cm diagonal.

$d = \sqrt{\ell^2 + w^2 + h^2}$	Formula for the diagonal of a right rectangular prism
$7 = \sqrt{4^2 + 3^2 + h^2}$	Substitute 7 for *d*, 4 for ℓ, and 3 for *w*.
$49 = 4^2 + 3^2 + h^2$	Square both sides of the equation.
$24 = h^2$	Simplify.
$4.9 \text{ cm} \approx h$	Take the square root of each side.

Find the number of vertices, edges, and faces of each polyhedron. Use your results to verify Euler's Formula.

1. 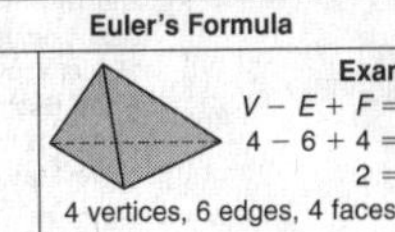

vertices: 8; edges: 12; faces: 6;
$8 - 12 + 6 = 2$

2. 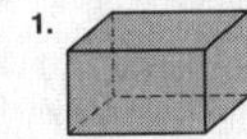

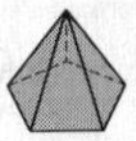

vertices: 6; edges: 10; faces: 6;
$6 - 10 + 6 = 2$

Find the unknown dimension in each figure. Round to the nearest tenth if necessary.

3. the length of the diagonal of a 6 cm by 8 cm by 11 cm rectangular prism

$d \approx 14.9$ cm

4. the height of a rectangular prism with a 4 in. by 5 in. base and a 9 in. diagonal

$h \approx 6.3$ in.

Copyright © by Holt, Rinehart and Winston. All rights reserved.

Copyright © by Holt, Rinehart and Winston. All rights reserved.

LESSON 10-3

Reteach

Formulas in Three Dimensions *continued*

A three-dimensional coordinate system has three perpendicular axes:

- *x*-axis
- *y*-axis
- *z*-axis

An *ordered triple* (x, y, z) is used to locate a point. The point at (3, 2, 4) is graphed at right.

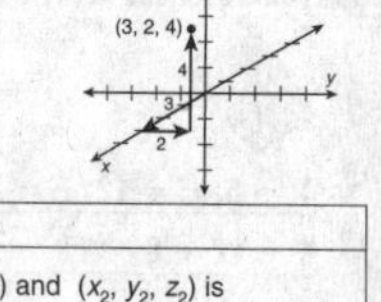

Formulas in Three Dimensions	
Distance Formula	The distance between the points (x_1, y_1, z_1) and (x_2, y_2, z_2) is $d = \sqrt{(x_2 - x_1)^2 + (y_2 - y_1)^2 + (z_2 - z_1)^2}$.
Midpoint Formula	The midpoint of the segment with endpoints (x_1, y_1, z_1) and (x_2, y_2, z_2) is $M\left(\frac{x_1 + x_2}{2}, \frac{y_1 + y_2}{2}, \frac{z_1 + z_2}{2}\right)$.

Find the distance between the points (4, 0, 1) and (2, 3, 0). Find the midpoint of the segment with the given endpoints.

$d = \sqrt{(x_2 - x_1)^2 + (y_2 - y_1)^2 + (z_2 - z_1)^2}$ Distance Formula

$= \sqrt{(2 - 4)^2 + (3 - 0)^2 + (0 - 1)^2}$ $(x_1, y_1, z_1) = (4, 0, 1), (x_2, y_2, z_2) = (2, 3, 0)$

$= \sqrt{4 + 9 + 1}$ Simplify.

$= \sqrt{14} \approx 3.7$ units Simplify.

The distance between the points (4, 0, 1) and (2, 3, 0) is about 3.7 units.

$M\left(\frac{x_1 + x_2}{2}, \frac{y_1 + y_2}{2}, \frac{z_1 + z_2}{2}\right) = M\left(\frac{4 + 2}{2}, \frac{0 + 3}{2}, \frac{1 + 0}{2}\right)$ Midpoint Formula

$= M(3, 1.5, 0.5)$ Simplify.

The midpoint of the segment with endpoints (4, 0, 1) and (2, 3, 0) is $M(3, 1.5, 0.5)$.

Find the distance between the given points. Find the midpoint of the segment with the given endpoints. Round to the nearest tenth if necessary.

5. (0, 0, 0) and (6, 8, 2)

   $d \approx 10.2$ units; $M(3, 4, 1)$

6. (0, 6, 0) and (4, 8, 0)

   $d \approx 4.5$ units; $M(2, 7, 0)$

7. (9, 1, 4) and (7, 0, 7)

   $d \approx 3.7$ units; $M(8, 0.5, 5.5)$

8. (2, 4, 1) and (3, 3, 5)

   $d \approx 4.2$ units; $M(2.5, 3.5, 3)$

Copyright © by Holt, Rinehart and Winston. All rights reserved.

LESSON 10-3

Challenge

Formulas in Three Dimensions

An Archimedean solid is a polyhedron whose faces are regular polygons (not necessarily of the same type) and whose polyhedral angles are all congruent. There are 13 such solids, of which only 5 are regular.

Euler's Formula states that for any polyhedron with *V* vertices, *E* edges, and *F* faces, $V - E + F = 2$.

This Archimedean solid is called the Great Rhombicosidodecahedron. The two-dimensional drawing is its net. A Great Rhombicosidodecahedron has 120 vertices and 180 edges.

The notation for the two-dimensional figures that form the faces of a polyhedron is f_3 for triangular faces, f_4 for quadrilateral faces, f_5 for pentagonal faces, and so on. The Great Rhombicosidodecahedron has 30 quadrilateral faces ($f_4 = 30$), 20 hexagonal faces ($f_6 = 20$), and 12 decagonal faces ($f_{10} = 12$).

1. How many faces does the Great Rhombicosidodecahedron have? **62**

Use the figure for Exercise 2. This Archimedean solid is called a Snub Dodecahedron. It has 150 edges and 92 faces. The faces are as follows: $f_3 = 80$ and $f_5 = 12$.

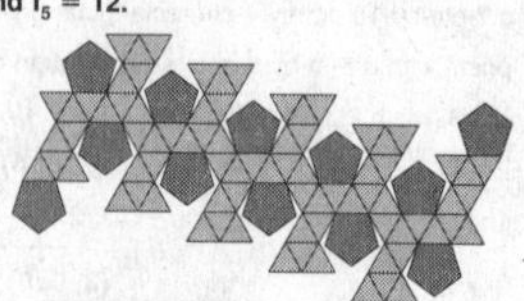

2. How many vertices does the Snub Dodecahedron have? **60**

Use the figure for Exercises 3–6. This Archimedean solid is called a Truncated Tetrahedron.

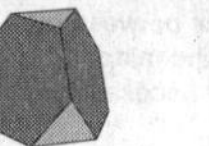

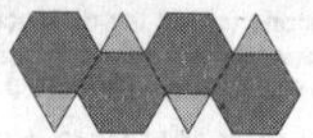

3. How many faces does the Truncated Tetrahedron have? **8**
4. How many edges does the Truncated Tetrahedron have? (*Hint:* Count all the sides of all the faces and divide by 2. Each edge consists of two sides touching.) **18**
5. How many vertices does the Truncated Tetrahedron have? **12**
6. Using proper notation, list the types of faces that are on a Truncated Tetrahedron and the number of each type. **$f_3 = 4$, $f_6 = 4$**

Copyright © by Holt, Rinehart and Winston. All rights reserved.

LESSON 10-3

Problem Solving

Formulas in Three Dimensions

1. What is the height of the rectangular prism? Round to the nearest tenth if necessary.

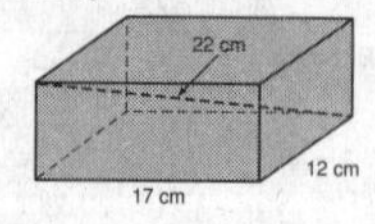

   7.1 cm

2. After lunch, Justin leaves the cafeteria to go to class, which is 22 feet north and 15 feet west of where he ate. The classroom is on the second floor, so it is 10 feet above the cafeteria. What is the actual distance between where Justin ate lunch and the classroom? Round to the nearest tenth.

   28.4 ft

3. Emily's hotel room is 18 feet south and 40 feet west of the pool. Her cousin Amber's hotel room is 22 feet north, 45 feet east, and 20 feet up on the third floor. How far apart are Emily's and Amber's rooms? Round to the nearest tenth.

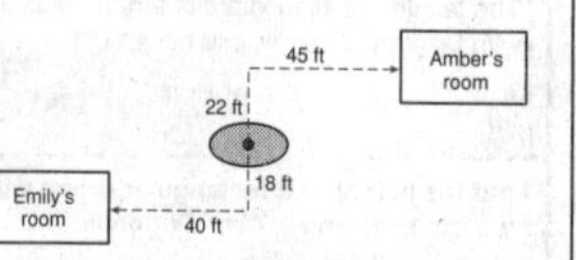

   96.0 ft

Choose the best answer.

4. How many faces, edges, and vertices does an octagonal pyramid have?
   - A 7 faces, 12 edges, 7 vertices
   - B 9 faces, 12 edges, 8 vertices
   - (C) 9 faces, 16 edges, 9 vertices
   - D 10 faces, 24 edges, 16 vertices

5. Which does NOT describe a polyhedron?
   - F 8 vertices, 12 edges, 6 faces
   - (G) 8 vertices, 10 edges, 6 faces
   - H 6 vertices, 9 edges, 5 faces
   - J 6 vertices, 10 edges, 6 faces

6. Point *R* has coordinates (8, 6, 1), and the midpoint of $\overline{RS}$ is $M(15, -2, 7)$. Which is the best estimate for the distance between point *R* and point *S*?
   - A 10.0 units
   - B 12.2 units
   - C 21.0 units
   - (D) 24.4 units

7. A rectangular prism has the following vertices. What is the volume of the prism?

   $A(0, 0, 4)$ $B(-4, 0, 0)$
   $C(-4, 2, 0)$ $D(0, 2, 0)$
   $E(0, 0, 0)$ $F(-4, 0, 4)$
   $G(-4, 2, 4)$ $H(0, 2, 4)$

   - F 4 units^3
   - G 16 units^3
   - (H) 32 units^3
   - J 64 units^3

Copyright © by Holt, Rinehart and Winston. All rights reserved.

LESSON 10-3

Reading Strategies

Use a Table

The table below shows some of the formulas used in three dimensions.

Formula	Diagram	Example
Length of Diagonal of a Right Rectangular Prism $d = \sqrt{\ell^2 + w^2 + h^2}$	5 cm, 4 cm, 6 cm, *d*	$d = \sqrt{6^2 + 4^2 + 5^2}$ $d = \sqrt{77} \approx 8.8$ cm
Distance Formula $d = \sqrt{(x_2 - x_1)^2 + (y_2 - y_1)^2 + (z_2 - z_1)^2}$	(1, 5, 6), (−7, 5, 0)	$d = \sqrt{(-7 - 1)^2 + (5 - 5)^2 + (0 - 6)^2}$ $d = \sqrt{(-8)^2 + (0)^2 + (-6)^2}$ $d = \sqrt{100} = 10$ units
Euler's Formula $V - E + F = 2$ V = number of vertices E = number of edges F = number of faces		Vertices: 5 Edges: 8 Faces: 5 $5 - 8 + 5 = 2$
Midpoint Formula $M = \left(\frac{x_1 + x_2}{2}, \frac{y_1 + y_2}{2}, \frac{z_1 + z_2}{2}\right)$	*M*, (−2, 3, 8), (6, −5, 10)	$\left(\frac{-2 + 6}{2}, \frac{3 + (-5)}{2}, \frac{8 + 10}{2}\right)$ $M = (2, -1, 9)$

Answer the following.

1. Write Euler's Formula in words.

   The number of vertices minus the number of edges plus the number of faces equals two.

2. Find the length of the diagonal of a 3 centimeter by 4 centimeter by 10 centimeter rectangular prism. Round to the nearest tenth.

   11.2 cm

Find the distance between the given points. Find the midpoint of the segment with the given points as endpoints. Round to the nearest tenth if necessary.

3. (2, 4, 5) and (6, 3, 1)

   $d =$ **5.7 units**

   $M =$ **$\left(4, \frac{7}{2}, 3\right)$**

4. (−1, 4, 7) and (5, 0, −5)

   $d =$ **14 units**

   $M =$ **(2, 2, 1)**

Copyright © by Holt, Rinehart and Winston. All rights reserved.

Copyright © by Holt, Rinehart and Winston. All rights reserved.

LESSON 10-4
Practice A
Surface Area of Prisms and Cylinders

Write each formula.

1. lateral area of a right prism with base perimeter P and height h — $L = Ph$
2. lateral area of a right cylinder with radius r and height h — $L = 2\pi rh$
3. surface area of a right prism with lateral area L and base area B — $S = L + 2B$
4. surface area of a cube with edge length s — $S = 6s^2$
5. surface area of a right cylinder with radius r and height h — $S = 2\pi rh + 2\pi r^2$

Find the lateral area and surface area of each right prism.

6. 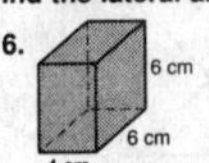

the rectangular prism
$L = 120\text{ cm}^2$; $S = 168\text{ cm}^2$

7. 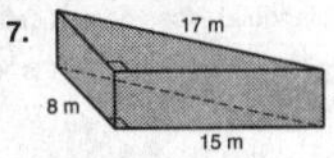 the triangular prism
$L = 160\text{ m}^2$; $S = 280\text{ m}^2$

8. a cube with edge length 2 ft — $L = 16\text{ ft}^2$; $S = 24\text{ ft}^2$

Find the lateral area and surface area of each right cylinder. Give your answers in terms of π.

9. 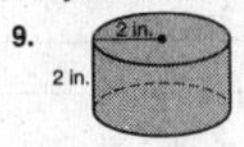

$L = 8\pi\text{ in}^2$; $S = 16\pi\text{ in}^2$

10. a cylinder with a radius of 3 mm and a height of 10 mm
$L = 60\pi\text{ mm}^2$; $S = 78\pi\text{ mm}^2$

A builder drills a hole through a cube of concrete, as shown in the figure. This cube will be an outlet for a water tap on the side of a house. Complete Exercises 11–14 to find the surface area of the figure. Round to the nearest tenth if necessary.

11. Find the surface area of the cube. — 384 in^2
12. Find the lateral area of the cylinder. — 50.3 in^2
13. Find twice the base area of the cylinder. — 6.3 in^2
14. The surface area of the figure is the surface area of the prism plus the lateral area of the cylinder minus twice the base area of the cylinder. Find the surface area of the figure. — 428.0 in^2

Copyright © by Holt, Rinehart and Winston. All rights reserved.

LESSON 10-4
Practice B
Surface Area of Prisms and Cylinders

Find the lateral area and surface area of each right prism. Round to the nearest tenth if necessary.

1. 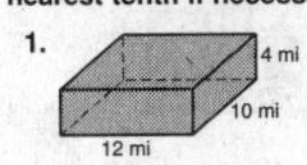

the rectangular prism
$L = 176\text{ mi}^2$; $S = 416\text{ mi}^2$

2. 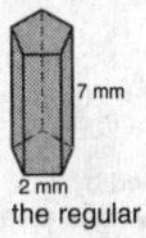 the regular pentagonal prism
$L = 70\text{ mm}^2$; $S = 83.8\text{ mm}^2$

3. a cube with edge length 20 inches — $L = 1600\text{ in}^2$; $S = 2400\text{ in}^2$

Find the lateral area and surface area of each right cylinder. Give your answers in terms of π.

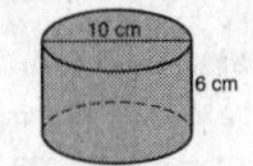

4. $L = 60\pi\text{ cm}^2$; $S = 110\pi\text{ cm}^2$
5. a cylinder with base area $169\pi\text{ ft}^2$ and a height twice the radius — $L = 676\pi\text{ ft}^2$; $S = 1014\pi\text{ ft}^2$
6. a cylinder with base circumference 8π m and a height one-fourth the radius — $L = 8\pi\text{ m}^2$; $S = 40\pi\text{ m}^2$

Find the surface area of each composite figure. Round to the nearest tenth.

7. 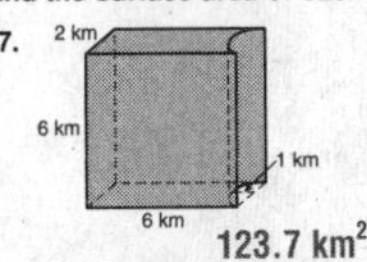

123.7 km^2

8. 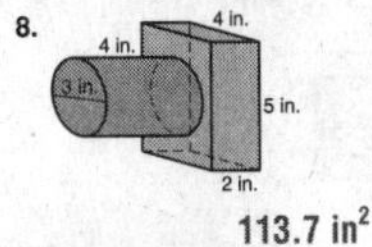

113.7 in^2

Describe the effect of each change on the surface area of the given figure.

9. 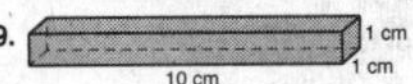

The dimensions are multiplied by 12.
The surface area is multiplied by 144.

10. 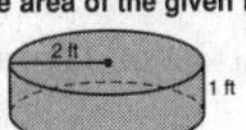

The dimensions are divided by 4.
The surface area is divided by 16.

Toby has eight cubes with edge length 1 inch. He can stack the cubes into three different rectangular prisms: 2-by-2-by-2, 8-by-1-by-1, and 2-by-4-by-1. Each prism has a volume of 8 cubic inches.

11. Tell which prism has the smallest surface-area-to-volume ratio. — 2-by-2-by-2
12. Tell which prism has the greatest surface-area-to-volume ratio. — 8-by-1-by-1

Copyright © by Holt, Rinehart and Winston. All rights reserved.

LESSON 10-4
Practice C
Surface Area of Prisms and Cylinders

A heat sink is a chunk of metal that draws unwanted heat away from delicate electronic components and releases the heat into the air. The figure shows a typical heat sink for a desktop computer processor chip. Each fin is 2 mm wide and is 4 mm from the next fin.

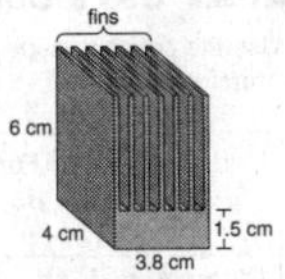

1. Find the surface area of the heat sink in square millimeters.
$31{,}840\text{ mm}^2$
2. Explain why the heat sink has fins. Possible answer: The fins greatly increase the surface area of the heat sink. The large surface area allows the heat to be radiated into the air more rapidly.

Find the surface area of each figure. Round to the nearest tenth if necessary.

3. 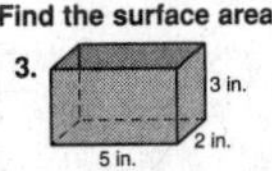
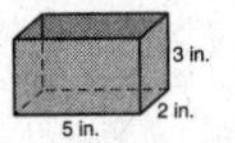

the rectangular prism with no top
52 in^2

4. 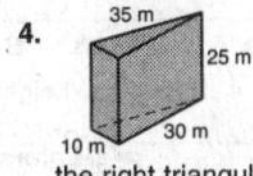

the right triangular prism
2153.1 m^2

5. 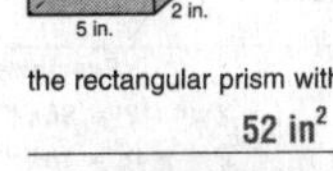

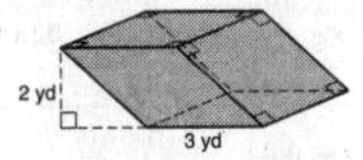

the oblique quadrilateral prism with each edge measuring 3 yards
48 yd^2

6. 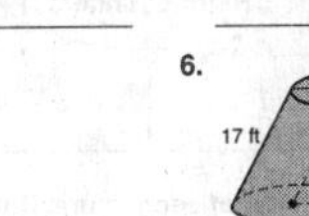

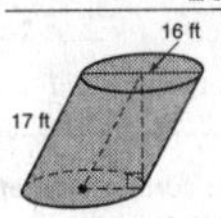

the oblique cylinder
1156.1 ft^2

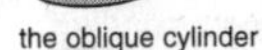

7. 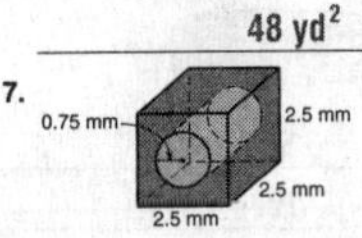
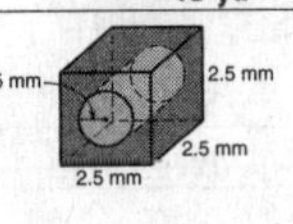

45.7 mm^2

8. 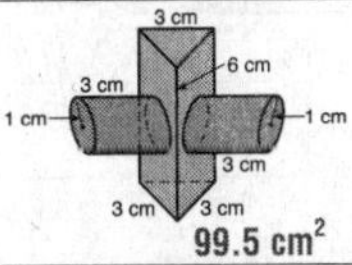 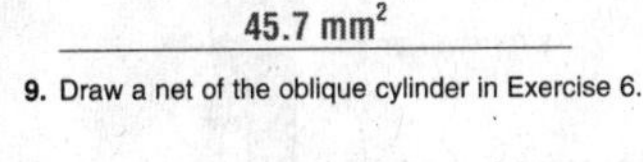
99.5 cm^2

9. Draw a net of the oblique cylinder in Exercise 6.

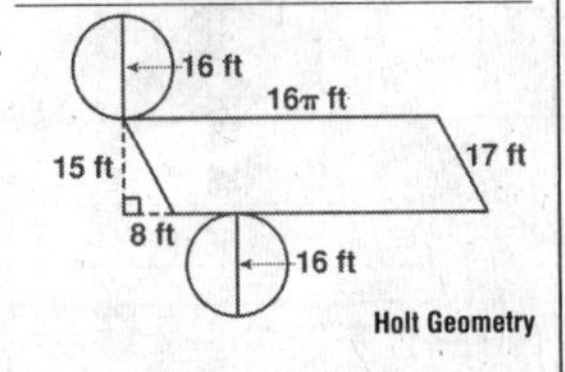

Copyright © by Holt, Rinehart and Winston. All rights reserved.

LESSON 10-4
Reteach
Surface Area of Prisms and Cylinders

The *lateral area* of a prism is the sum of the areas of all the *lateral faces.* A lateral face is not a base. The **surface area** is the total area of all faces.

Lateral and Surface Area of a Right Prism		
Lateral Area	The lateral area of a right prism with base perimeter P and height h is $L = Ph$.	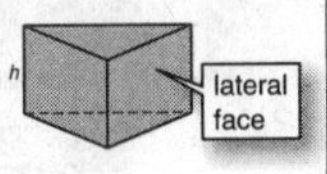 lateral face
Surface Area	The surface area of a right prism with lateral area L and base area B is $S = L + 2B$, or $S = Ph + 2B$.	

The lateral area of a right cylinder is the curved surface that connects the two bases. The **surface area** is the total area of the curved surface and the bases.

Lateral and Surface Area of a Right Cylinder		
Lateral Area	The lateral area of a right cylinder with radius r and height h is $L = 2\pi rh$.	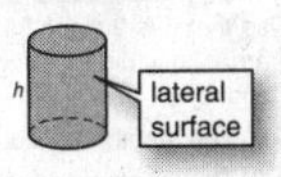 lateral surface
Surface Area	The surface area of a right cylinder with lateral area L and base area B is $S = L + 2B$, or $S = 2\pi rh + 2\pi r^2$.	

Find the lateral area and surface area of each right prism.

1. 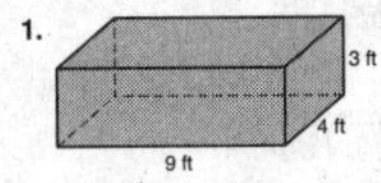

$L = 78\text{ ft}^2$; $S = 150\text{ ft}^2$

2. 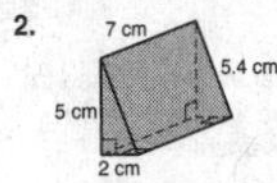

$L = 86.8\text{ cm}^2$; $S = 96.8\text{ cm}^2$

Find the lateral area and surface area of each right cylinder. Give your answers in terms of π.

3. 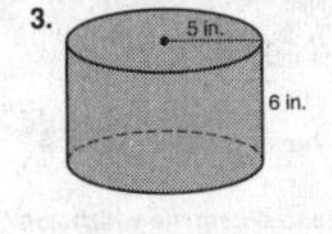

$L = 60\pi\text{ in}^2$; $S = 110\pi\text{ in}^2$

4. 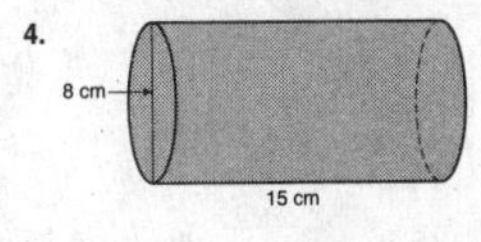

$L = 120\pi\text{ cm}^2$; $S = 152\pi\text{ cm}^2$

Copyright © by Holt, Rinehart and Winston. All rights reserved.

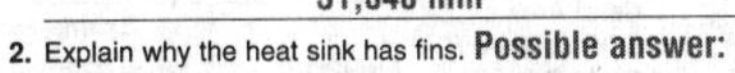
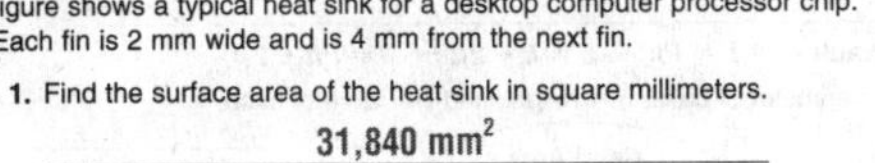
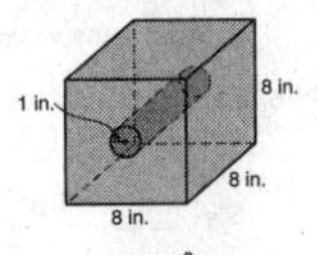

Copyright © by Holt, Rinehart and Winston. All rights reserved.

LESSON 10-4 Reteach
Surface Area of Prisms and Cylinders *continued*

You can find the surface area of a composite three-dimensional figure like the one shown at right.

surface area of small prism + surface area of large prism − hidden surfaces

The dimensions are multiplied by 3. Describe the effect on the surface area.

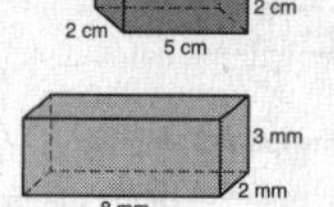

original surface area:		new surface area, dimensions multiplied by 3:	
$S = Ph + 2B$		$S = Ph + 2B$	
$= 20(3) + 2(16)$	$P = 20, h = 3, B = 16$	$= 60(9) + 2(144)$	$P = 60, h = 9, B = 144$
$= 92$ mm^2	Simplify.	$= 828$ mm^2	Simplify.

Notice that $92 \cdot 9 = 828$. If the dimensions are multiplied by 3, the surface area is multiplied by 3^2, or 9.

Find the surface area of each composite figure. Be sure to subtract the hidden surfaces of each part of the composite solid. Round to the nearest tenth.

5. 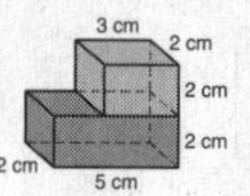

$S = 68$ cm^2

6.
$S \approx 64.6$ in^2

Describe the effect of each change on the surface area of the given figure.

7. The length, width, and height are multiplied by 2.

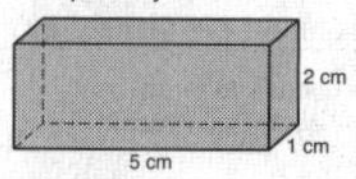

The surface area is multiplied by 4.

8. The height and radius are multiplied by $\frac{1}{2}$.

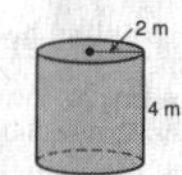

The surface area is multiplied by $\frac{1}{4}$.

Copyright © by Holt, Rinehart and Winston. All rights reserved.

LESSON 10-4 Challenge
Surface Area and Volume of Semiregular Polyhedra

A **semiregular polyhedron** is a convex polyhedron whose faces are bounded by two or more types of regular polygons in such a way that the arrangement of polygons at each vertex of the polyhedron is identical.

The figure at right is a semiregular polyhedron called a *cuboctahedron.* Its faces are bounded by equilateral triangles and squares. You can think of it as the figure obtained if you "cut off" eight congruent pieces from a cube in the manner shown.

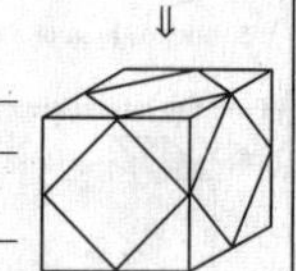

1. a. How many square faces does a cuboctahedron have? **6**
   b. How many triangular faces does a cuboctahedron have? **8**
2. What type of figure is each piece that is "cut off" from the original cube? **pyramid**

Suppose that the length of each edge of a cuboctahedron is 10 inches.

3. a. What is the area of each square face? **100 in^2**
   b. What is the area of each triangular face? **$25\sqrt{3}$ in^2**
   c. What is the total surface area? **$(600 + 200\sqrt{3})$ in^2**
4. a. What is the length of each edge of the original cube from which the cuboctahedron was "cut"? **$10\sqrt{2}$ in.**
   b. What is the volume of this cube? **$2000\sqrt{2}$ in^3**
   c. What is the volume of each piece that was "cut off" from the cube? **$\frac{125}{3}\sqrt{2}$ in^3**
   d. What is the volume of the cuboctahedron? **$\frac{5000}{3}\sqrt{2}$ in^3**
5. Generalize your results from Exercises 3 and 4 to write formulas for the surface area S and volume V of a cuboctahedron with edge of length n.

   $S =$ **$6n^2 + 2n^2\sqrt{3}$** $V =$ **$\frac{5}{3}n^3\sqrt{2}$**
6. When eight congruent pieces are cut from a cube in the manner shown at right, the result is a semiregular polyhedron called a *truncated cube.* Write formulas for the surface area S and volume V of a truncated cube with edge of length m.

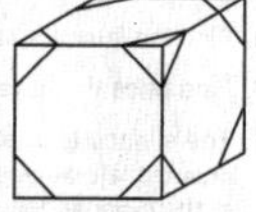

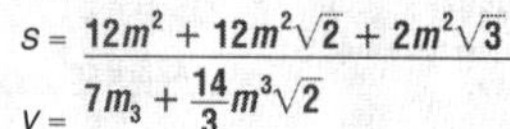

   $S =$ **$12m^2 + 12m^2\sqrt{2} + 2m^2\sqrt{3}$**

   $V =$ **$7m_3 + \frac{14}{3}m^3\sqrt{2}$**

Copyright © by Holt, Rinehart and Winston. All rights reserved.

LESSON 10-4 Problem Solving
Surface Area of Prisms and Cylinders

1. The lateral area of the regular pentagonal prism below is 220 mm^2. What is the surface area? Round to the nearest tenth if necessary.

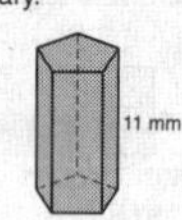

275.1 mm^2

2. A sheet of metal 8 feet long and 6 feet wide is to be cut into cylindrical cans like the one shown. How many lateral surfaces for the cans can be cut from the metal with as little waste as possible?

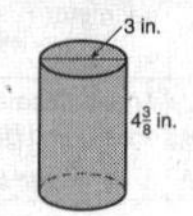

160 lateral surfaces

Choose the best answer.

3. The surface area of a cube is increased so that it is 9 times its original surface area. How did the length of the cube change?
   A The length was doubled.
   (B) The length was tripled.
   C The length was quadrupled.
   D The length was multiplied by 9.
4. A rectangular prism has a surface area of 152 square inches. If the length, width, and height are all changed to $\frac{1}{2}$ their original size, what will be the new surface area of the prism?
   F 19 in^2
   (G) 38 in^2
   H 76 in^2
   J 114 in^2
5. Determine the surface area exposed to the air of the composite figure shown. Round to the nearest tenth.

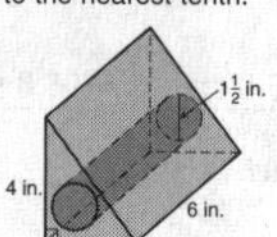

   A 98.1 in^2
   B 107.6 in^2
   (C) 108.7 in^2
   D 110.5 in^2
6. Which of the two cylindrical cans has a greater surface area?

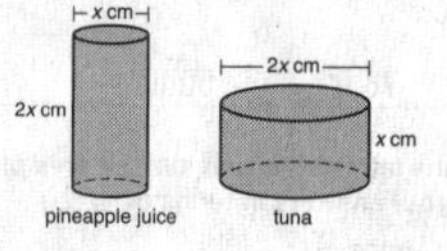

   F pineapple juice can
   (G) tuna can
   H The two cans have the same surface area.
   J It is impossible to determine which can has a greater surface area.

Copyright © by Holt, Rinehart and Winston. All rights reserved.

LESSON 10-4 Reading Strategies
Use a Concept Map

Use the concept maps below to help you understand and use lateral and surface area formulas.

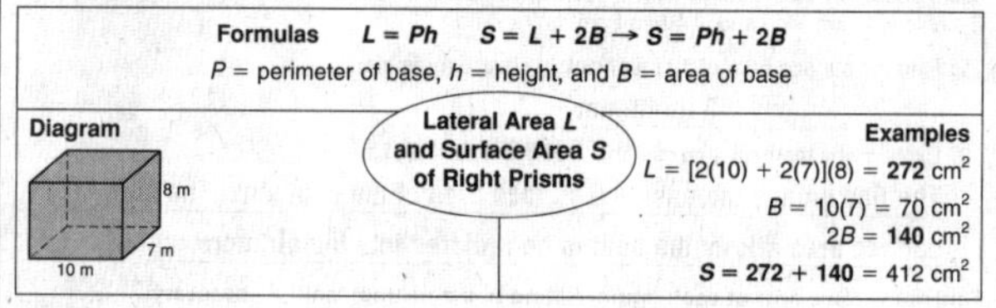

Formulas $L = Ph$ $S = L + 2B \rightarrow S = Ph + 2B$
P = perimeter of base, h = height, and B = area of base
Lateral Area L and Surface Area S of Right Prisms
Diagram
8 m
7 m
10 m
Examples
$L = [2(10) + 2(7)](8) = 272$ cm^2
$B = 10(7) = 70$ cm^2
$2B = 140$ cm^2
$S = 272 + 140 = 412$ cm^2

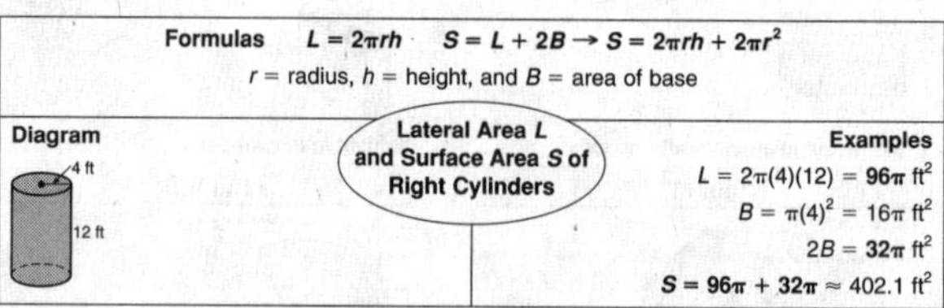

Formulas $L = 2\pi rh$ $S = L + 2B \rightarrow S = 2\pi rh + 2\pi r^2$
r = radius, h = height, and B = area of base
Lateral Area L and Surface Area S of Right Cylinders
Diagram
4 ft
12 ft
Examples
$L = 2\pi(4)(12) = 96\pi$ ft^2
$B = \pi(4)^2 = 16\pi$ ft^2
$2B = 32\pi$ ft^2
$S = 96\pi + 32\pi \approx 402.1$ ft^2

Find the lateral area and surface area of each figure. Round to the nearest tenth if necessary.

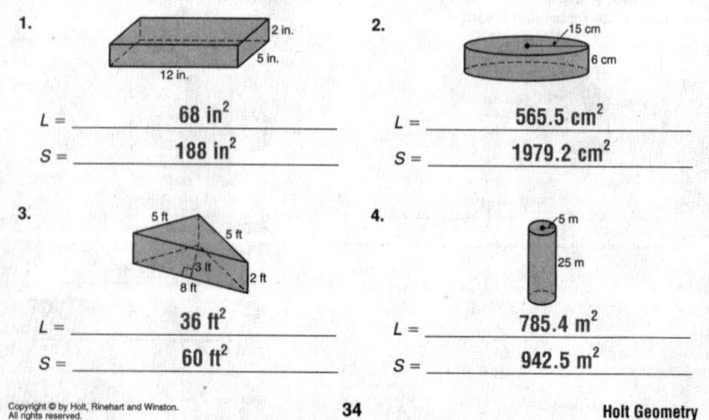

1. $L =$ **68 in^2** $S =$ **188 in^2**
2. $L =$ **565.5 cm^2** $S =$ **1979.2 cm^2**
3. $L =$ **36 ft^2** $S =$ **60 ft^2**
4. $L =$ **785.4 m^2** $S =$ **942.5 m^2**

Copyright © by Holt, Rinehart and Winston. All rights reserved.

Copyright © by Holt, Rinehart and Winston. All rights reserved.

LESSON 10-5

Practice A

Surface Area of Pyramids and Cones

Write each formula.

1. lateral area of a regular pyramid with base perimeter P and slant height ℓ — $L = \frac{1}{2}P\ell$
2. lateral area of a right cone with radius r and slant height ℓ — $L = \pi r\ell$
3. surface area of a regular pyramid with lateral area L and base area B — $S = L + B$
4. surface area of a right cone with lateral area L and base area B — $S = L + B$

Find the lateral area and surface area of each regular pyramid. Round to the nearest tenth if necessary.

5. 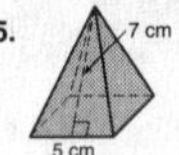

the regular square pyramid

$L = 70\text{ cm}^2$; $S = 95\text{ cm}^2$

6. 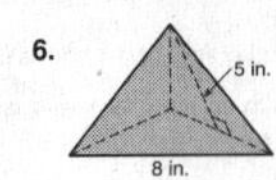

the regular triangular pyramid

$L = 60\text{ in}^2$; $S = 87.7\text{ in}^2$

Find the lateral area and surface area of each right cone. Give your answers in terms of π.

7. 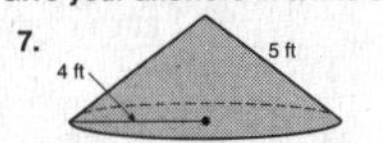

$L = 20\pi\text{ ft}^2$; $S = 36\pi\text{ ft}^2$

8. a right cone with radius 3 m and slant height 12 m

$L = 36\pi\text{ m}^2$; $S = 45\pi\text{ m}^2$

Complete Exercises 9–11 to describe the effect on the surface area of dividing the dimensions of a cone by 2. Give your answers in terms of π.

9. Find the surface area of a right cone with radius 2 yards and slant height 6 yards. — $16\pi\text{ yd}^2$
10. Find the surface area of a right cone with radius 1 yard and slant height 3 yards. — $4\pi\text{ yd}^2$
11. Describe the effect on the surface area of dividing the dimensions of a right cone by 2.

The surface area is divided by 4.

12. Find the surface area of the composite figure in terms of π.

$124\pi\text{ mm}^2$

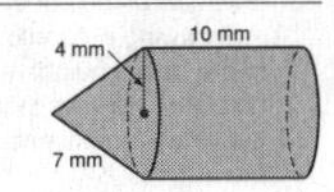

Copyright © by Holt, Rinehart and Winston. All rights reserved.

LESSON 10-5

Practice B

Surface Area of Pyramids and Cones

Find the lateral area and surface area of each regular right solid. Round to the nearest tenth if necessary.

1. 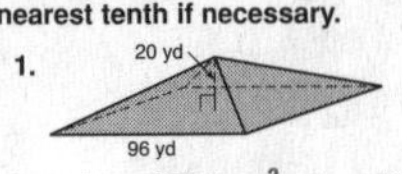

$L = 9984\text{ yd}^2$; $S = 19{,}200\text{ yd}^2$

2. 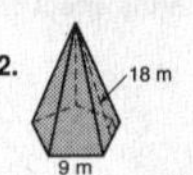

$L = 405\text{ m}^2$; $S = 544.4\text{ m}^2$

3. a regular hexagonal pyramid with base edge length 12 mi and slant height 15 mi

$L = 540\text{ mi}^2$; $S = 914.1\text{ mi}^2$

Find the lateral area and surface area of each right cone. Give your answers in terms of π.

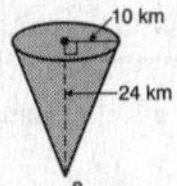

4. $L = 260\pi\text{ km}^2$; $S = 360\pi\text{ km}^2$ (10 km, 24 km)
5. a right cone with base circumference 14π ft and slant height 3 times the radius — $L = 147\pi\text{ ft}^2$; $S = 196\pi\text{ ft}^2$
6. a right cone with diameter 240 cm and altitude 35 cm

$L = 15{,}000\pi\text{ cm}^2$; $S = 29{,}400\pi\text{ cm}^2$

Describe the effect of each change on the surface area of the given figure.

7. 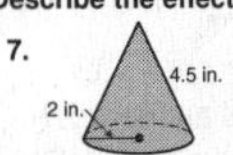

The dimensions are multiplied by $\frac{1}{5}$.

The surface area is multiplied by $\frac{1}{25}$.

8. 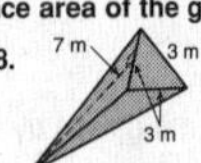

The dimensions are multiplied by $\frac{3}{2}$.

The surface area is multiplied by $\frac{9}{4}$.

Find the surface area of each composite figure. Round to the nearest tenth if necessary.

9. 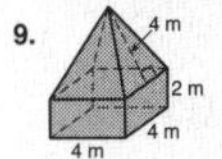

$S = 80\text{ m}^2$

10. 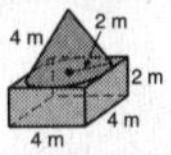 (4 m, 2 m, 2 m, 4 m, 4 m)

$S = 76.6\text{ m}^2$

11. The water cooler at Mohammed's office has small conical paper cups for drinking. He uncurls one of the cups and measures the paper. Based on the diagram of the uncurled cup, find the diameter of the cone.

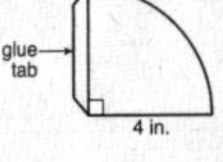

$d = 2\text{ in.}$

Copyright © by Holt, Rinehart and Winston. All rights reserved.

LESSON 10-5

Practice C

Surface Area of Pyramids and Cones

Use the figure for Exercises 1–3. The figure shown can be curled into an open-topped cone. (16 cm)

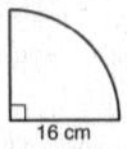

1. Find the radius of the top of the cone. — 4 cm
2. Find the radius of a circle that has the same area as the lateral area of the cone. — 8 cm
3. Name the mathematical relationship between the answer to Exercise 2 and the radius and slant height of the cone. — geometric mean

Find the radius of a circle that has the same area as the lateral area of each cone described in Exercises 4–6. Give exact answers.

4. $\ell = 32$ ft; $r = 8$ ft — 16 ft
5. $\ell = 68$ m; $r = 17$ m — 34 m
6. $\ell = 5$ cm; $r = 2$ cm — $\sqrt{10}$ cm

7. A cone with an open base can be formed from any partial circle. Develop a formula for the lateral area of a cone based on ℓ, the slant height, and d, the degree measure of the interior angle in the partial circle. (For instance, d in the figure above equals 90.) — $L = \frac{\pi d\ell^2}{360}$

The figures below can be curled into open frustums. Find the lateral area of the outside and the radius of both bases in each frustum. Round to the nearest tenth.

8. 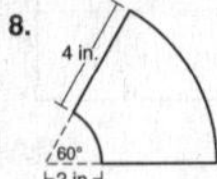

$L = 16.8\text{ in}^2$; $r_1 = 0.3$ in.; $r_2 = 1$ in.

9. 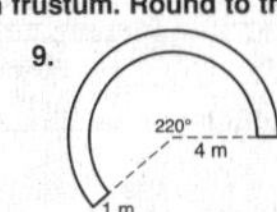

$L = 17.3\text{ m}^2$; $r_1 = 2.4$ m; $r_2 = 3.1$ m

10. Terry is driving through the desert when she notices the engine is low on oil. She has a few quarts of motor oil in the trunk of the car, but she does not have a funnel. Fortunately, Terry finds a piece of $8\frac{1}{2}$-by-11-inch notebook paper in the backseat. She wants to cut the paper to make a funnel with a 1-inch diameter hole on the bottom and the longest slant height possible. Find the diameter of the top of the funnel. Draw the pattern Terry will cut out before curling up the funnel. — $5\frac{1}{2}$ in.

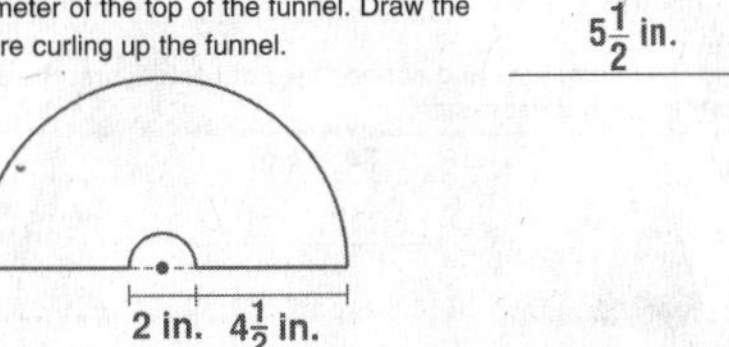

Copyright © by Holt, Rinehart and Winston. All rights reserved.

LESSON 10-5

Reteach

Surface Area of Pyramids and Cones

Lateral and Surface Area of a Regular Pyramid		
Lateral Area	The lateral area of a regular pyramid with perimeter P and slant height ℓ is $L = \frac{1}{2}P\ell$.	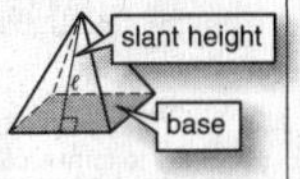 slant height; base
Surface Area	The surface area of a regular pyramid with lateral area L and base area B is $S = L + B$, or $S = \frac{1}{2}P\ell + B$.	

Lateral and Surface Area of a Right Cone		
Lateral Area	The lateral area of a right cone with radius r and slant height ℓ is $L = \pi r\ell$.	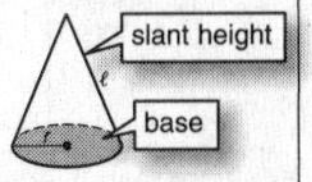 slant height; base
Surface Area	The surface area of a right cone with lateral area L and base area B is $S = L + B$, or $S = \pi r\ell + \pi r^2$.	

Find the lateral area and surface area of each regular pyramid. Round to the nearest tenth.

1. 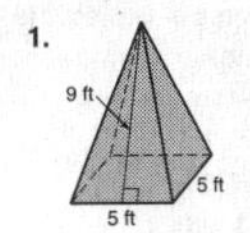

$L = 90\text{ ft}^2$; $S = 115\text{ ft}^2$

2. 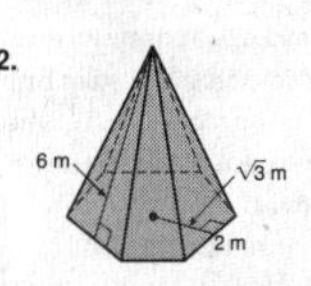

$L = 36\text{ m}^2$; $S \approx 46.4\text{ m}^2$

Find the lateral area and surface area of each right cone. Give your answers in terms of π.

3. 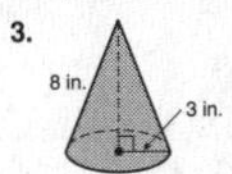

$L = 24\pi\text{ in}^2$; $S = 33\pi\text{ in}^2$

4. 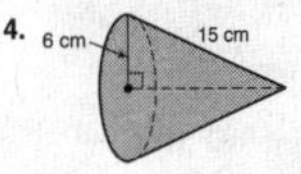

$L = 90\pi\text{ cm}^2$; $S = 126\pi\text{ cm}^2$

Copyright © by Holt, Rinehart and Winston. All rights reserved.

Copyright © by Holt, Rinehart and Winston. All rights reserved.

LESSON 10-5 **Reteach**

Surface Area of Pyramids and Cones continued

The radius and slant height of the cone at right are doubled. Describe the effect on the surface area.

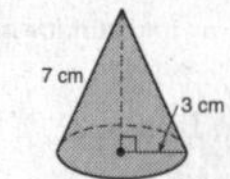

original surface area:

$S = \pi r\ell + \pi r^2$

$= \pi(3)(7) + \pi(3)^2$ $r = 3, \ell = 7$

$= 30\pi$ cm^2 Simplify.

new surface area, dimensions doubled:

$S = \pi r\ell + \pi r^2$

$= \pi(6)(14) + \pi(6)^2$ $r = 6, \ell = 14$

$= 120\pi$ cm^2 Simplify.

If the dimensions are doubled, then the surface area is multiplied by 2^2, or 4.

Describe the effect of each change on the surface area of the given figure.

5. The dimensions are tripled.

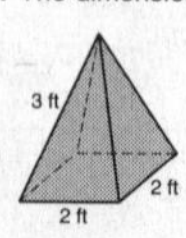

The surface area is multiplied by 9.

6. The dimensions are multiplied by $\frac{1}{2}$.

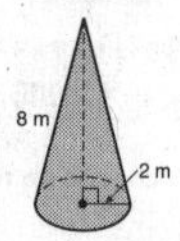

The surface area is multiplied by $\frac{1}{4}$.

Find the surface area of each composite figure.

7. *Hint:* Do not include the base area of the pyramid or the upper surface area of the rectangular prism.

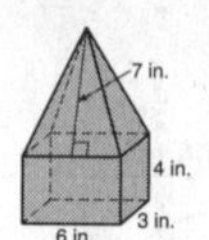

$S = 153$ in^2

8. *Hint:* Add the lateral areas of the cones.

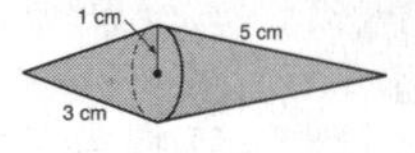

$S = 8\pi$ cm^2

Copyright © by Holt, Rinehart and Winston. All rights reserved.

LESSON 10-5 **Challenge**

Making Nets for Right Cones

Suppose that you need to make a model of a cone that has the dimensions given in the figure at right. You know that the net for the cone consists of a circular region for the base and a region bounded by a sector of a circle for the lateral surface. But how do you know the exact size of each piece?

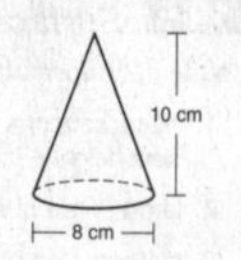

Give each measure for the right cone shown at right. When necessary, round to the nearest tenth of a centimeter.

1. radius 4 cm
2. circumference 25.1 cm
3. height 10 cm
4. slant height 10.8 cm
5. A sketch of a net for the cone shown above is given at right.
   a. Label the sketch with as many of the measures from Exercises 1–4 as possible.
   b. Suppose that you were to use the sketch to draw the net. Which important measure is still needed? m∠ABC

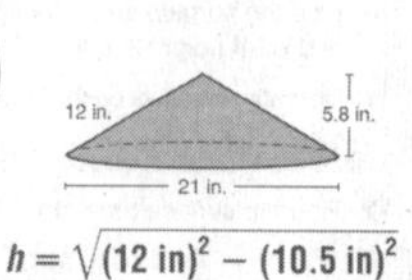

6. Refer to the net for the cone that you labeled in Exercise 5.
   a. Suppose that the *entire* circle with center at point *B* was drawn. What would be its circumference? 67.7 cm
   b. What is the length of the arc that is drawn from *A* to *C*? 25.1 cm
   c. What percent of the entire circle is the arc from *A* to *C*? $\frac{4}{10.77} \approx 37.1\%$
   d. Multiply 360° by your percent from part **c**. What is the measure of ∠*ABC*, rounded to the nearest whole degree? 134°
7. Refer to your results from Exercises 5 and 6. Using a compass, ruler, and protractor, draw an accurate real-size net for the cone. Then assemble the net to make a model of the cone. Check students' work.
8. A sketch of a net for a right cone is given at right. In the blank space to its right, draw the cone, making the height and diameter of the cone in the drawing proportional to the actual height and diameter. Be sure to label the height and diameter.

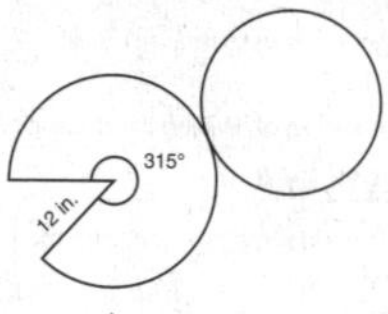

$h = \sqrt{(12 \text{ in})^2 - (10.5 \text{ in})^2}$

$h = 5.8$ in.

Copyright © by Holt, Rinehart and Winston. All rights reserved.

LESSON 10-5 **Problem Solving**

Surface Area of Pyramids and Cones

1. Find the diameter of a right cone with slant height 18 centimeters and surface area 208π square centimeters. 16 cm
2. Find the surface area of a regular pentagonal pyramid with base area 49 square meters and slant height 13 meters. Round to the nearest tenth. 222.4 mm^2
3. A piece of paper in the shape shown is folded to form a cone. What is the diameter of the base of the cone that is formed? Round to the nearest tenth.

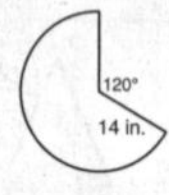

18.7 in.

4. The right cone has a surface area of 240π square millimeters. What is the radius of the cone?

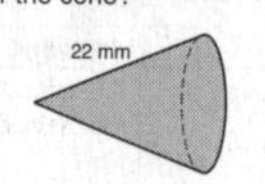

8 mm

Choose the best answer.

5. A square pyramid has a base with a side length of 9 centimeters and a slant height that is 4 centimeters more than $1\frac{1}{2}$ times the length of the base. Find the surface area of the pyramid.
   A 162 cm^2 C 315 cm^2
   B 243 cm^2 **(D) 396 cm^2**
6. A cone has a surface area of 64π square inches. If the radius and height are each multiplied by $\frac{3}{4}$, what will be the new surface area of the cone?
   (F) 36π in^2 H 60π in^2
   G 48π in^2 J 96π in^2
7. Find the surface area of the composite figure. Round to the nearest tenth.

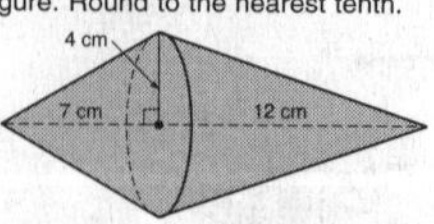

   A 238.8 cm^2 C 311.0 cm^2
   (B) 260.3 cm^2 D 361.3 cm^2
8. A cone has a base diameter of 6 yards. What is the slant height of the cone if it has the same surface area as the square pyramid shown? Round to the nearest tenth.

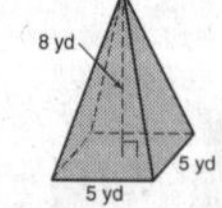

   F 8.1 yd H 11.3 yd
   (G) 8.5 yd J 25.6 yd

Copyright © by Holt, Rinehart and Winston. All rights reserved.

LESSON 10-5 **Reading Strategies**

Compare and Contrast

The diagram below summarizes the similarities and differences between regular pyramids and right cones and their lateral and surface areas.

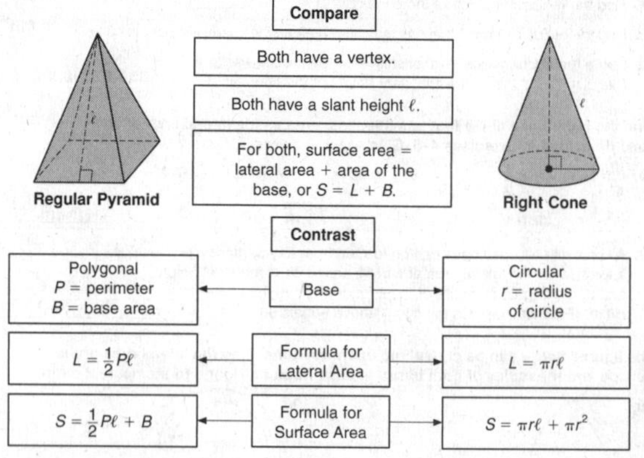

Answer the following.

1. Look at the pyramid and cone above. Why do you think the slant height is so named?
   Possible answer: it is not perpendicular to the base, is slanted diagonally, and is the height of a lateral face
2. Look at the formulas for surface area for each figure. Why do you think the formula for the pyramid uses *B* for area of the base and the formula for the cone does not?
   The base of a pyramid could be different polygons. The base of a cone is always a circle.

Find the lateral area and surface area of each figure. Round to the nearest tenth if necessary.

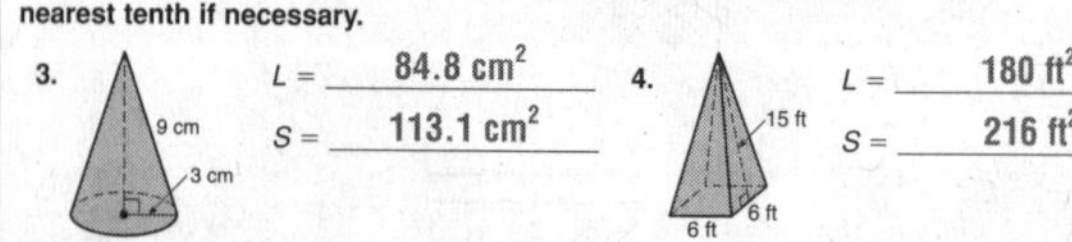

3. $L =$ 84.8 cm^2 $S =$ 113.1 cm^2
4. $L =$ 180 ft^2 $S =$ 216 ft^2

Copyright © by Holt, Rinehart and Winston. All rights reserved.

Copyright © by Holt, Rinehart and Winston. All rights reserved.

LESSON 10-6

Practice A

Volume of Prisms and Cylinders

Write each formula.

1. volume of a cube with edge length s — $V = s^3$
2. volume of a prism with base area B and height h — $V = Bh$
3. volume of a cylinder with radius r and height h — $V = \pi r^2 h$
4. volume of a right rectangular prism with length ℓ, width w, and height h — $V = \ell wh$

Find the volume of each prism. Round to the nearest tenth if necessary.

5. 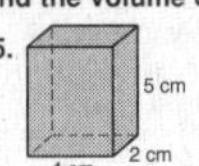

the right rectangular prism

$V = 40 \text{ cm}^3$

6. 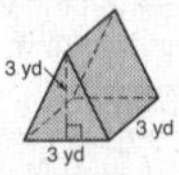 the triangular prism

$V \approx 13.5 \text{ yd}^3$

7. Laetitia needs to store 8 boxes while she is moving. Each box is a cube with edge length 3 feet. A storage facility charges $0.75 for every cubic foot of storage per month. Find the amount of money Laetitia will pay to store her boxes for one month. — **$162**

Find the volume of each cylinder. Give your answers both in terms of π and rounded to the nearest tenth.

8. 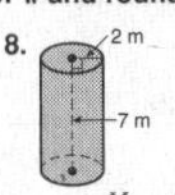

$V = 28\pi \text{ m}^3$; $V \approx 88.0 \text{ m}^3$

9. a cylinder with diameter 20 in. and height 2 in.

$V = 200\pi \text{ in}^3$; $V \approx 628.3 \text{ in}^3$

Complete Exercises 10–12 to describe the effect on the volume of multiplying each dimension of a prism by 3.

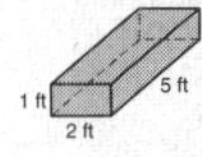

10. Find the volume of the prism. — $V = 10 \text{ ft}^3$
11. Find the volume of the prism after each dimension is multiplied by 3. — $V = 270 \text{ ft}^3$
12. Describe the effect on the volume of multiplying each dimension of a prism by 3.

The volume is multiplied by 27.

13. Find the volume of the composite figure. Round to the nearest tenth.

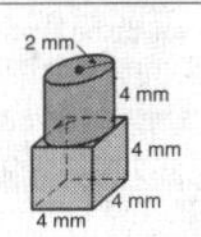

$V \approx 114.3 \text{ mm}^3$

Copyright © by Holt, Rinehart and Winston. All rights reserved.

LESSON 10-6

Practice B

Volume of Prisms and Cylinders

Find the volume of each prism. Round to the nearest tenth if necessary.

1. 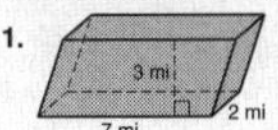

the oblique rectangular prism

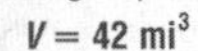

$V = 42 \text{ mi}^3$

2. 

the regular octagonal prism

$V \approx 7242.6 \text{ mm}^3$

3. a cube with edge length 0.75 m — $V \approx 0.4 \text{ m}^3$

Find the volume of each cylinder. Give your answers both in terms of π and rounded to the nearest tenth.

4. 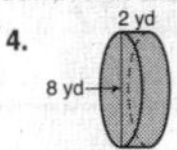

$V = 32\pi \text{ yd}^3$; $V \approx 100.5 \text{ yd}^3$

5. 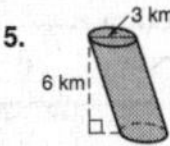

$V = 13.5\pi \text{ km}^3$; $V \approx 42.4 \text{ km}^3$

6. a cylinder with base circumference 18π ft and height 10 ft — $V = 810\pi \text{ ft}^3$; $V \approx 2544.7 \text{ ft}^3$

7. CDs have the dimensions shown in the figure. Each CD is 1 mm thick. Find the volume in cubic centimeters of a stack of 25 CDs. Round to the nearest tenth.

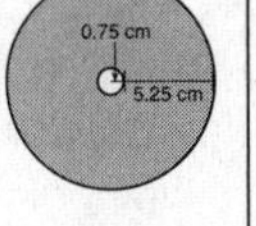

$V \approx 278.3 \text{ cm}^3$

Describe the effect of each change on the volume of the given figure.

8. 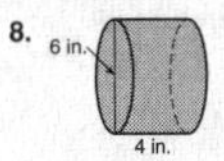

The dimensions are halved.

The volume is divided by 8.

9. 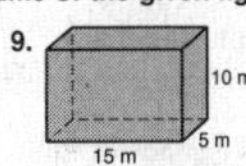

The dimensions are divided by 5.

The volume is divided by 125.

Find the volume of each composite figure. Round to the nearest tenth.

10. 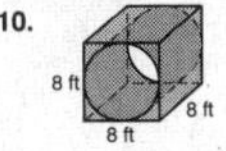

$V \approx 109.9 \text{ ft}^3$

11. 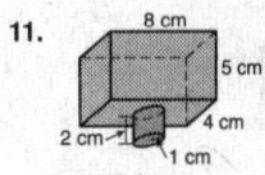

$V \approx 166.3 \text{ cm}^3$

Copyright © by Holt, Rinehart and Winston. All rights reserved.

LESSON 10-6

Practice C

Volume of Prisms and Cylinders

1. Find the volume-to-surface-area ratio for these two cylinders. Round to the nearest tenth.

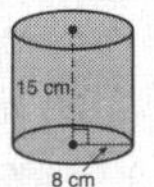

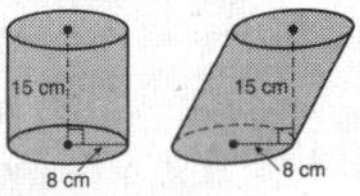

2.6; 2.6

A chocolate bar is in the shape of a rectangular prism with length 5 in., width $2\frac{1}{4}$ in., and height $\frac{1}{4}$ in. The bar weighs 1.75 ounces. The chart shows some of the nutritional information for the chocolate bar. Round your answers to Exercises 2–4 to the nearest hundredth.

Serving size $\frac{1}{2}$ bar
Calories 135
Total fat 8 g (12% DV)
Total carb 14 g (5% DV)

2. Find the density of the chocolate bar (ounces/cubic inch). — 0.62 oz/in^3
3. Find the volume of chocolate that contains 100 calories. — 1.04 in^3
4. The "% DV" indicates the percentage of the recommended daily amount for that nutrient. Find the volume of chocolate that would provide 100% of the recommended daily amount of carbohydrates. (*Note:* This is NOT a healthy diet.) — 28.13 in^3

In the sciences, quantities of liquids are measured in liters and milliliters. One milliliter of water has the same volume as a cube with edge length 1 cm.

5. Tell what size cube has the same volume as 1 liter of water.

a cube with edge length 10 cm

6. In a science lab, liquids are often measured out in tall, thin cylinders called graduated cylinders. One graduated cylinder has a diameter of 2 centimeters, and 8 milliliters of water are poured into it. Tell how high the water will reach. Round to the nearest tenth. — 2.5 cm

Find the volume of each figure. Round to the nearest tenth.

7. 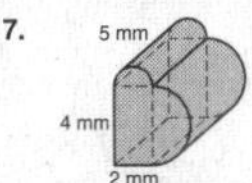

$V = 79.3 \text{ mm}^3$

8. 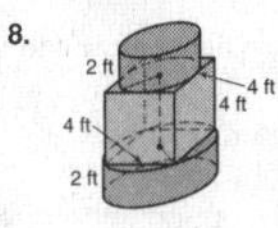

$V = 139.4 \text{ ft}^3$

Copyright © by Holt, Rinehart and Winston. All rights reserved.

LESSON 10-6

Reteach

Volume of Prisms and Cylinders

Volume of Prisms		
Prism	The volume of a prism with base area B and height h is $V = Bh$.	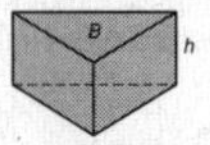
Right Rectangular Prism	The volume of a right rectangular prism with length ℓ, width w, and height h is $V = \ell wh$.	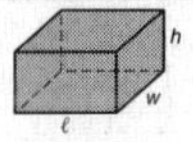
Cube	The volume of a cube with edge length s is $V = s^3$.	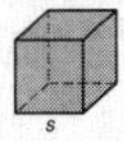

Volume of a Cylinder

The volume of a cylinder with base area B, radius r, and height h is

$V = Bh$, or $V = \pi r^2 h$.

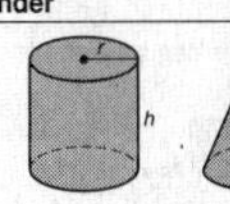

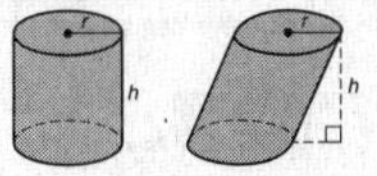

Find the volume of each prism.

1. 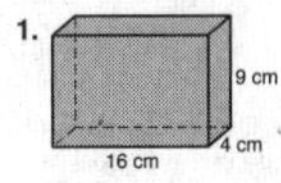

$V = 576 \text{ cm}^3$

2. 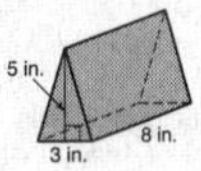 $V = 60 \text{ in}^3$

Find the volume of each cylinder. Give your answers both in terms of π and rounded to the nearest tenth.

3. 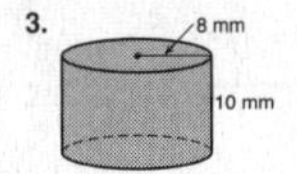

$V = 640\pi \text{ mm}^3 \approx 2010.6 \text{ mm}^3$

4. 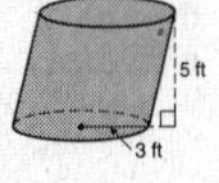

$V = 45\pi \text{ ft}^3 \approx 141.4 \text{ ft}^3$

Copyright © by Holt, Rinehart and Winston. All rights reserved.

Copyright © by Holt, Rinehart and Winston. All rights reserved.

LESSON 10-6

Reteach

Volume of Prisms and Cylinders continued

The dimensions of the prism are multiplied by $\frac{1}{3}$. Describe the effect on the volume.

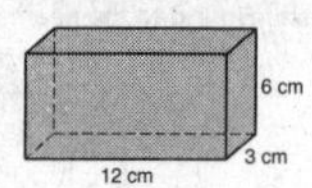

original volume:		new volume, dimensions multiplied by $\frac{1}{3}$:	
$V = \ell wh$		$V = \ell wh$	
$= (12)(3)(6)$	$\ell = 12, w = 3, h = 6$	$= (4)(1)(2)$	$\ell = 4, w = 1, h = 2$
$= 216 \text{ cm}^3$	Simplify.	$= 8 \text{ cm}^3$	Simplify.

Notice that $216 \cdot \frac{1}{27} = 8$. If the dimensions are multiplied by $\frac{1}{3}$, the volume is multiplied by $\left(\frac{1}{3}\right)^3$, or $\frac{1}{27}$.

Describe the effect of each change on the volume of the given figure.

5. The dimensions are multiplied by 2.

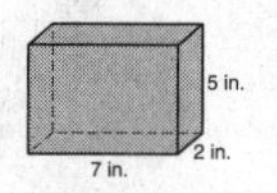

The volume is multiplied by 8.

6. The dimensions are multiplied by $\frac{1}{4}$.

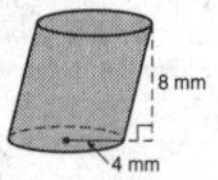

The volume is multiplied by $\frac{1}{64}$.

Find the volume of each composite figure. Round to the nearest tenth.

7.

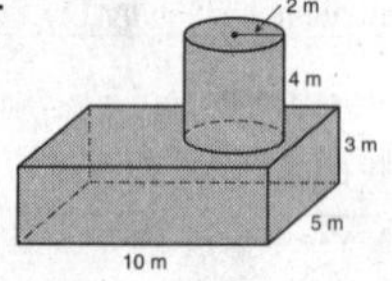

$V \approx 200.3 \text{ m}^3$

8.

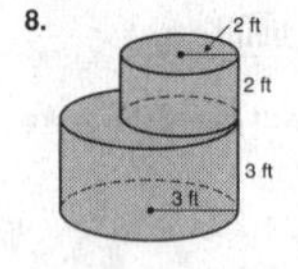

$V \approx 110.0 \text{ ft}^3$

Copyright © by Holt, Rinehart and Winston. All rights reserved.

LESSON 10-6

Challenge

Using the Volume Formula to Adjust a Recipe

Most baking recipes specify a certain size of baking pan. When that size of pan is not available, you may be able to adjust the recipe to a different size. Since many items are baked in the shape of a rectangular prism, this adjustment can be done by calculating volumes with the formula $V = \ell wh$. For example, the recipe at right requires a 13 × 9 × 2-inch pan. This type of pan is shaped like a rectangular prism that is 13 inches long, 9 inches wide, and 2 inches high, as shown below.

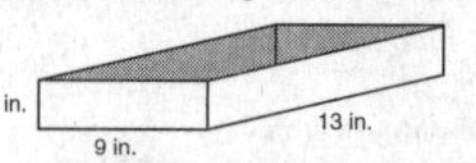

Crackle Bars

3 tablespoons margarine
40 regular marshmallows, or
4 cups miniature marshmallows
6 cups toasted rice cereal

Melt the margarine in a large saucepan over low heat. Add the marshmallows and stir until they are completely melted. Remove from heat. Stir in the cereal until it is coated. Press the mixture into a greased 13 × 9 × 2-inch pan. Cut bars when cool.

Refer to the recipe for Crackle Bars that is given above. Assume that when you prepare the mixture according to the recipe, it fills the 13 × 9 × 2-inch pan to an unknown height of *h* inches.

1. What is the volume of the recipe mixture? **$117h \text{ in}^3$**
2. Suppose that you only have an 8 × 8 × 2-inch pan. What would be the volume of the mixture if the pan were filled to a height of *h* inches? **$64h \text{ in}^3$**
3. What percent of the recipe mixture would fill the 8 × 8 × 2-inch pan to a height of *h* inches? Round to the nearest whole percent. **55%**
4. Calculate the amount of each ingredient needed to make enough mixture to fill the 8 × 8 × 2-inch pan to a height of *h* inches with no extra mixture.
   a. margarine (1 tablespoon equals 3 teaspoons) **$1\frac{2}{3}$ tablespoons**
   b. regular marshmallows **22**
   c. miniature marshmallows **$2\frac{1}{5}$ cups**
   d. toasted rice cereal **$3\frac{1}{3}$ cups**
5. Use the method from Exercises 1–4. Adjust the amounts of ingredients in the Crackle Bar recipe so that the mixture fills a pan of the given dimensions to a height of *h* inches. When necessary, round to reasonable measures. Write your answers on a separate sheet of paper. **Check students' work.**
   a. 9 in. × 9 in. × 2 in.
   b. 15 in. × 10 in. × $1\frac{1}{2}$ in.
   c. 25 cm × 35 cm × 4 cm
6. Explain how to adjust the Crackle Bar recipe so the mixture fills a pan that is *a* inches long, *b* inches wide, and *c* inches high to a height of 2*h* inches.
   Multiply the amount of each ingredient by $\frac{2ab}{117}$. Round to reasonable measures. Make sure that *c* is less than or equal to 2*h*.

Copyright © by Holt, Rinehart and Winston. All rights reserved.

LESSON 10-6

Problem Solving

Volume of Prisms and Cylinders

1. A cylindrical juice container has the dimensions shown. About how many cups of juice does this container hold? (*Hint:* 1 cup ≈ 14.44 in³)

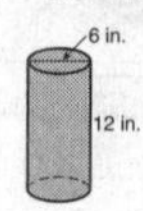

about 23.50 cups

2. A large cylindrical cooler is $2\frac{1}{2}$ feet high and has a diameter of $1\frac{1}{2}$ feet. It is filled $\frac{3}{4}$ high with water for athletes to use during their soccer game. Estimate the volume of the water in the cooler in gallons. (*Hint:* 1 gallon ≈ 231 in³)

about 25 gal

Choose the best answer.

3. How many 3-inch cubes can be placed inside the box?

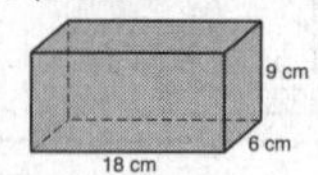

A 27
(B) 36
C 45
D 72

4. A cylinder has a volume of $4\pi \text{ cm}^3$. If the radius and height are each tripled, what will be the new volume of the cylinder?

F $12\pi \text{ cm}^3$
G $36\pi \text{ cm}^3$
H $64\pi \text{ cm}^3$
(J) $108\pi \text{ cm}^3$

5. What is the volume of the composite figure with the dimensions shown in the three views? Round to the nearest tenth.

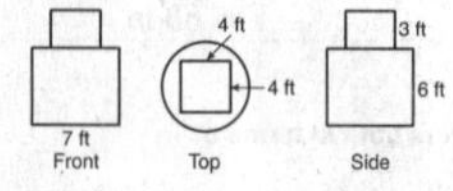

A 182.9 ft³
B 205.7 ft³
(C) 278.9 ft³
D 971.6 ft³

6. Find the expression that can be used to determine the volume of the composite figure shown.

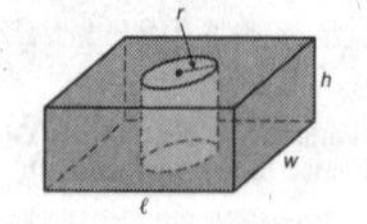

(F) $\ell wh - \pi r^2 h$
G $\pi r^2 h + \ell wh$
H $\pi r^2 h - \ell wh$
J $\ell wh + 2\pi r^2 h$

Copyright © by Holt, Rinehart and Winston. All rights reserved.

LESSON 10-6

Reading Strategies

Use a Table

The tables below show formulas for finding the volume of different three-dimensional figures.

Cylinders	Diagram	Formula	Example
Oblique and right cylinders	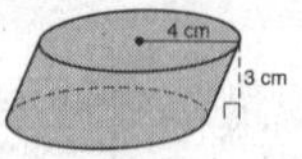 4 cm, 3 cm	$V = Bh$ where *B* is the area of the base, so $V = \pi r^2 h$	$V = \pi(4^2)(3) \text{ cm}^3$ $= \pi(48) \text{ cm}^3$ $\approx 150.8 \text{ cm}^3$

Prisms	Diagram	Formula	Example
Cube	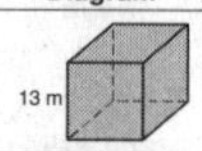 13 m	$V = s^3$	$V = 13^3 \text{ m}^3$ $= 2197 \text{ m}^3$
Rectangular prism	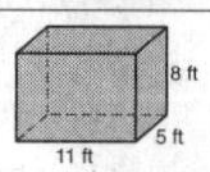 8 ft, 5 ft, 11 ft	$V = \ell wh$	$V = (11)(5)(8) \text{ ft}^3$ $= 440 \text{ ft}^3$
Other oblique and right prisms	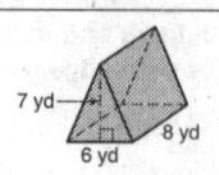 7 yd, 8 yd, 6 yd	$V = Bh$ where *B* is the area of the base	$B = \frac{1}{2}(6)(7) \text{ yd}^2 = 21 \text{ yd}^2$ $V = 21(8) \text{ yd}^3 = 168 \text{ yd}^3$

Answer the following. Round to the nearest tenth if necessary.

1. Find the volume of a cube with edge length 8 centimters. **$V = 512 \text{ cm}^3$**
2. Find the volume of a cylinder with height 15 inches and radius 3 inches. **$V \approx 424.1 \text{ in}^3$**

Find the volume of each figure. Round to the nearest tenth if necessary.

3.

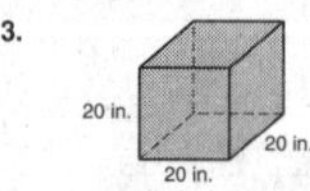

$V = 8000 \text{ in}^3$

4. **$V \approx 785.4 \text{ ft}^3$**

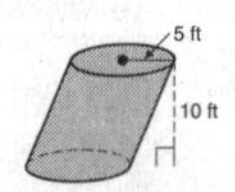

5. **$V = 144 \text{ m}^3$**

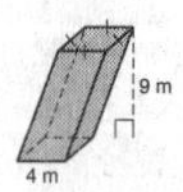

Copyright © by Holt, Rinehart and Winston. All rights reserved.

Copyright © by Holt, Rinehart and Winston. All rights reserved.

LESSON 10-7

Practice A

Volume of Pyramids and Cones

Write each formula.

1. volume of a pyramid with base area B and height h $V = \frac{1}{3}Bh$
2. volume of a cone with radius r and height h $V = \frac{1}{3}\pi r^2 h$

Find the volume of each pyramid.

3. 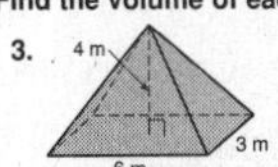

the rectangular pyramid
$V = 24\text{ m}^3$

4. 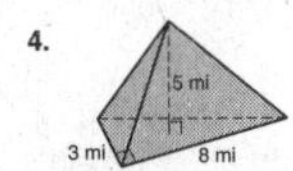

the right triangular pyramid
$V = 20\text{ mi}^3$

5. a square pyramid with side length 10 in. and height 12 in. $V = 400\text{ in}^3$

Find the volume of each cone. Give your answers both in terms of π and rounded to the nearest tenth.

6. 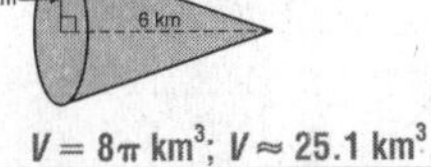

$V = 8\pi\text{ km}^3$; $V \approx 25.1\text{ km}^3$

7. a cone with diameter 15 yd and height 10 yd
$V = 187.5\pi\text{ yd}^3$; $V \approx 589.0\text{ yd}^3$

8. An ant lion is an insect that digs cone-shaped pits in loose dirt to trap ants. When an ant tumbles down into the pit, the ant lion eats it. A typical ant lion pit has a radius of 1 inch and a depth of 2 inches. Find the volume of dirt the ant lion moved to dig its hole. Round to the nearest tenth.
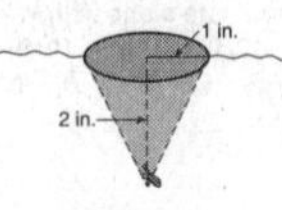

$V \approx 2.1\text{ in}^3$

Complete Exercises 9–11 to describe the effect on the volume of dividing the dimensions of the cone by 3. Give your answers in terms of π.

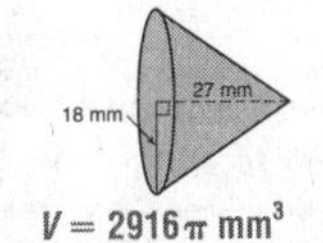

9. Find the volume of the cone. $V = 2916\pi\text{ mm}^3$
10. Find the volume of the cone after the radius and height are divided by 3. $V = 108\pi\text{ mm}^3$
11. Describe the effect on the volume after dividing the dimensions of a cone by 3. The volume is divided by 27.
12. Find the volume of the composite figure.
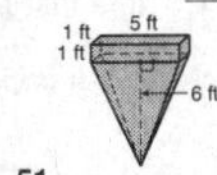

$V = 15\text{ ft}^3$

Copyright © by Holt, Rinehart and Winston. All rights reserved.

LESSON 10-7

Practice B

Volume of Pyramids and Cones

Find the volume of each pyramid. Round to the nearest tenth if necessary.

1. 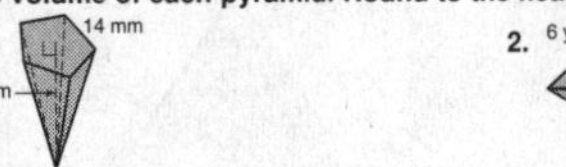

the regular pentagonal pyramid
$V \approx 3934.2\text{ mm}^3$

2. 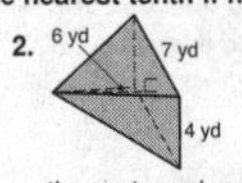

the rectangular right pyramid
$V = 56\text{ yd}^3$

3. Giza in Egypt is the site of the three great Egyptian pyramids. Each pyramid has a square base. The largest pyramid was built for Khufu. When first built, it had base edges of 754 feet and a height of 481 feet. Over the centuries, some of the stone eroded away and some was taken for newer buildings. Khufu's pyramid today has base edges of 745 feet and a height of 471 feet. To the nearest cubic foot, find the difference between the original and current volumes of the pyramid.
$4{,}013{,}140\text{ ft}^3$

Find the volume of each cone. Give your answers both in terms of π and rounded to the nearest tenth.

4. 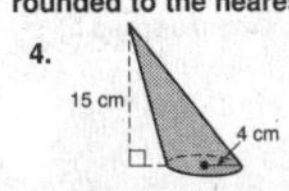

$V = 80\pi\text{ cm}^3$; $V \approx 251.3\text{ cm}^3$

5. 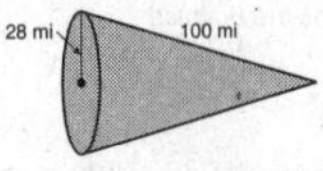 28 mi, 100 mi
$V = 25{,}088\pi\text{ mi}^3$; $V \approx 78{,}816.3\text{ mi}^3$

6. a cone with base circumference 6π m and a height equal to half the radius
$V = 4.5\pi\text{ m}^3$; $V \approx 14.1\text{ m}^3$

7. Compare the volume of a cone and the volume of a cylinder with equal height and base area.
The volume of the cone is one-third the volume of the cylinder.

Describe the effect of each change on the volume of the given figure.

8. 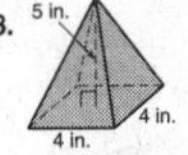

The dimensions are multiplied by $\frac{2}{3}$. The volume is multiplied by $\frac{8}{27}$.

9. 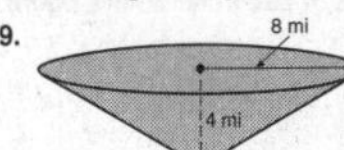

The dimensions are tripled. The volume is multiplied by 27.

Find the volume of each composite figure. Round to the nearest tenth.

10. 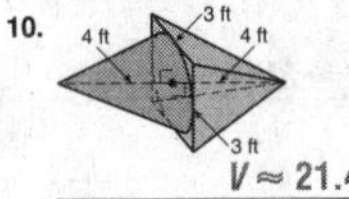

$V \approx 21.4\text{ ft}^3$

11. 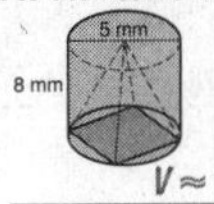 5 mm, 8 mm
$V \approx 123.7\text{ mm}^3$

Copyright © by Holt, Rinehart and Winston. All rights reserved.

LESSON 10-7

Practice C

Volume of Pyramids and Cones

1. The figure shows a square-based pyramid with a height equal to the side length of the base. The segment connecting the vertex to its closest corner is perpendicular to the base. Draw a net for this pyramid below. Then, using a separate sheet of paper, cut out three of these shapes. Fold them into pyramids, and assemble them into a cube. Describe what this demonstrates.
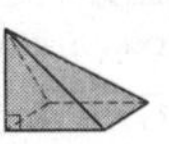
Possible answer: A square pyramid with height equal to an edge length has one-third the volume of a cube with the same edge length.
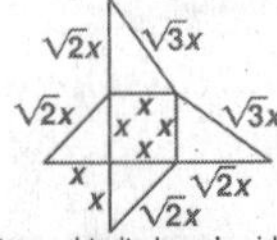

2. A square pyramid has a height equal to its base's side length, and its surface area is equal to its volume (although the units are different). Find the side length of the base. Give both an exact answer and an answer rounded to the nearest tenth. $3 + 3\sqrt{5}$; 9.7
3. A cone has a height equal to its radius, and its surface area is equal to its volume (although the units are different). Find the radius. Give both an exact answer and an answer rounded to the nearest tenth. $3 + 3\sqrt{2}$; 7.2
4. Draw a figure that has exactly two-thirds the volume of this regular hexagonal prism.
Possible answer:
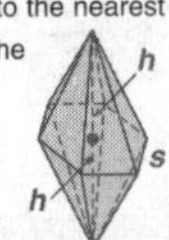

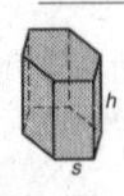

Find the volume of each figure. Round to the nearest tenth if necessary.

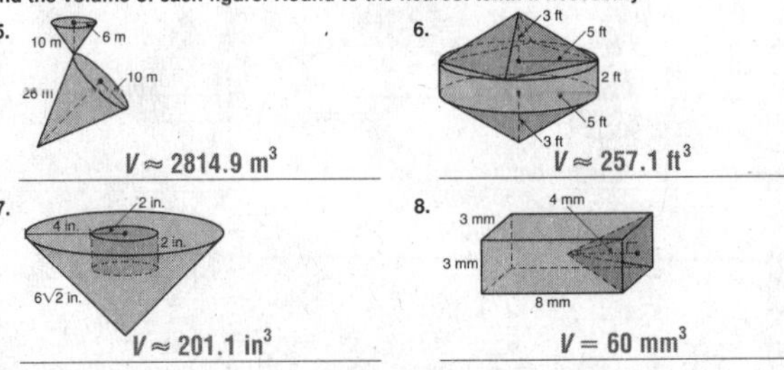

5. $V \approx 2814.9\text{ m}^3$
6. $V \approx 257.1\text{ ft}^3$
7. $V \approx 201.1\text{ in}^3$
8. $V = 60\text{ mm}^3$

Copyright © by Holt, Rinehart and Winston. All rights reserved.

LESSON 10-7

Reteach

Volume of Pyramids and Cones

Volume of a Pyramid

The volume of a pyramid with base area B and height h is
$V = \frac{1}{3}Bh$.

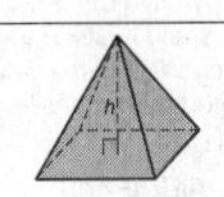

Volume of a Cone

The volume of a cone with base area B, radius r, and height h is
$V = \frac{1}{3}Bh$, or $V = \frac{1}{3}\pi r^2 h$.

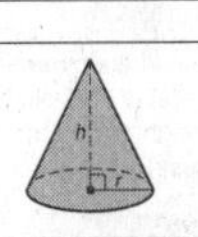

Find the volume of each pyramid. Round to the nearest tenth if necessary.

1. 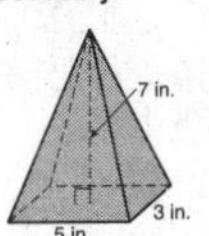

$V = 35\text{ in}^3$

2. 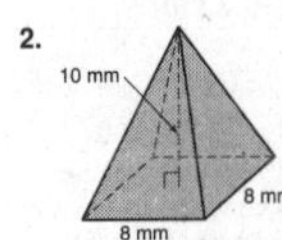

$V \approx 213.3\text{ mm}^3$

Find the volume of each cone. Give your answers both in terms of π and rounded to the nearest tenth.

3. 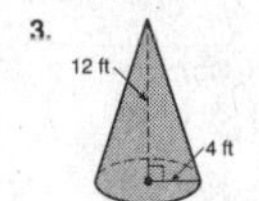

$V = 64\pi\text{ ft}^3 \approx 201.1\text{ ft}^3$

4. 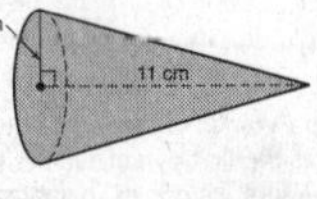 3 cm, 11 cm
$V = 33\pi\text{ cm}^3 \approx 103.7\text{ cm}^3$

Copyright © by Holt, Rinehart and Winston. All rights reserved.

Copyright © by Holt, Rinehart and Winston. All rights reserved.

LESSON 10-7
Reteach
Volume of Pyramids and Cones continued

The radius and height of the cone are multiplied by $\frac{1}{2}$. Describe the effect on the volume.

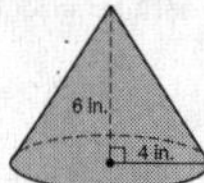

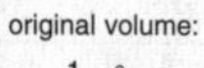

original volume:		new volume, dimensions multiplied by $\frac{1}{2}$:	
$V = \frac{1}{3}\pi r^2 h$		$V = \frac{1}{3}\pi r^2 h$	
$= \frac{1}{3}\pi(4)^2(6)$	$r = 4, h = 6$	$= \frac{1}{3}\pi(2)^2(3)$	$r = 2, h = 3$
$= 32\pi \text{ in}^3$	Simplify.	$= 4\pi \text{ in}^3$	Simplify.

If the dimensions are multiplied by $\frac{1}{2}$, then the volume is multiplied by $\left(\frac{1}{2}\right)^3$, or $\frac{1}{8}$.

Describe the effect of each change on the volume of the given figure.

5. The dimensions are doubled.

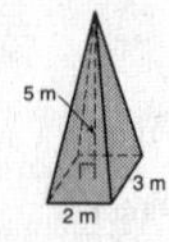

The volume is multiplied by 8.

6. The radius and height are multiplied by $\frac{1}{3}$.

6 ft
18 ft

The volume is multiplied by $\frac{1}{27}$.

Find the volume of each composite figure. Round to the nearest tenth if necessary.

7.

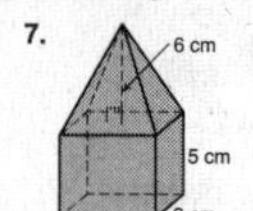

$V = 126 \text{ cm}^3$

8.

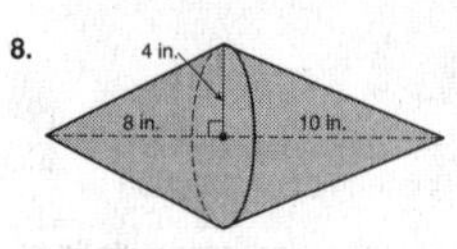

$V \approx 301.6 \text{ in}^3$

Copyright © by Holt, Rinehart and Winston. All rights reserved.

LESSON 10-7
Challenge
Volume of Pyramids and Cones

Draw a figure in a three-dimensional coordinate plane with vertices $A(0, 0, 0)$, $B(8, 0, 0)$, $C(0, 10, 0)$, $D(8, 10, 0)$, and $E(4, 4, 12)$. The height of the figure is EF. F is at $(4, 4, 0)$.

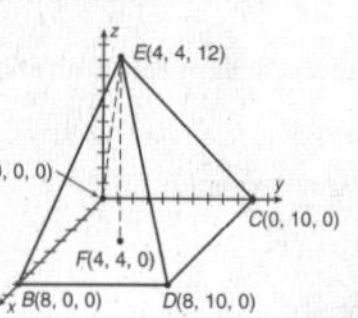

1. Name the base of this figure and its shape. **rectangle *ABDC***
2. What type of figure is *ABDCE*? **rectangular pyramid**
3. What is the formula for finding the volume of this figure? $V = \frac{1}{3}Bh$
4. Use the distance formula to find *EF*. **12 units**
5. Find *AB*. **8 units**
6. Find *AC*. **10 units**
7. Find the volume of the figure. $V = 320 \text{ units}^3$

Draw a figure in a three dimensional coordinate plane with vertices $P(0, 0, 0)$, $N(-10, 0, 0)$, $L(0, -10, 0)$, $M(-10, -10, 0)$, $K(-5, -5, -6)$, $J(-5, -5, 7)$.

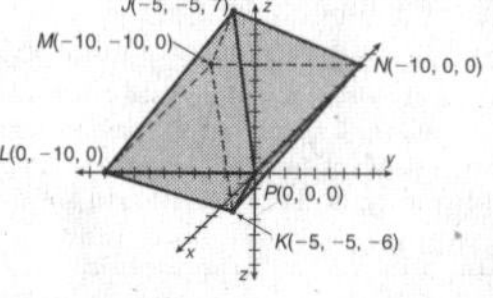

8. Name the base of this figure and its shape. You might want to plot the base coordinates on an *x-y* plane. **square *LMNP***
9. Find *LP*. **10 units**
10. Find the area of the base in Exercise 8. 100 units^2
11. What type of figure is *JLMNPK*? **octahedron**
12. What is the formula for finding the volume of this figure? Explain.
    Consider the octahedron as two square pyramids with different altitudes, h_1 and h_2. $V = \frac{1}{3}B(h_1 + h_2)$ Note that altitude is always a positive number.
13. Find the volume of the three-dimensional shape named in Exercise 11. Round to the nearest tenth. $V \approx 433.3 \text{ units}^3$

Copyright © by Holt, Rinehart and Winston. All rights reserved.

LESSON 10-7
Problem Solving
Volume of Pyramids and Cones

1. A regular square pyramid has a base area of 196 meters and a lateral area of 448 square meters. What is the volume of the pyramid? Round your answer to the nearest tenth.
   $V \approx 940.0 \text{ m}^3$

2. A paper cone for serving roasted almonds has a volume of 406π cubic centimeters. A smaller cone has half the radius and half the height of the first cone. What is the volume of the smaller cone? Give your answer in terms of π.
   $V = 50.75\pi \text{ cm}^3$

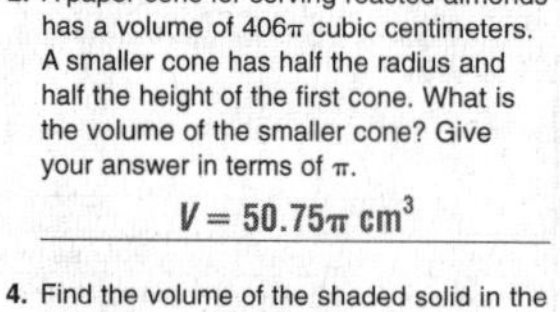

3. The hexagonal base in the pyramid is a regular polygon. What is the volume of the pyramid if its height is 9 centimeters? Round to the nearest tenth.

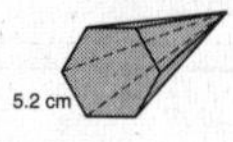

$V \approx 210.8 \text{ cm}^3$

4. Find the volume of the shaded solid in the figure shown. Give your answer in terms of π.

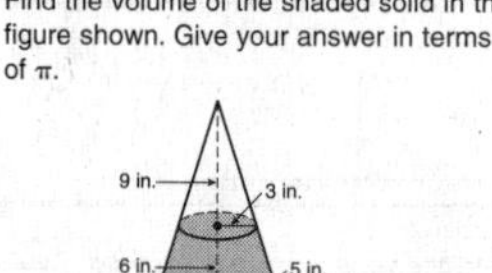

$V = 98\pi \text{ in}^3$

Choose the best answer.

5. The diameter of the cone equals the width of the cube, and the figures have the same height. Find the expression that can be used to determine the volume of the composite figure.

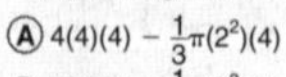

Ⓐ $4(4)(4) - \frac{1}{3}\pi(2^2)(4)$
B $4(4)(4) + \frac{1}{3}\pi(2^2)(4)$
C $4(4)(4) - \pi(2^2)(4)$
D $4(4)(4) + \frac{1}{3}\pi(2^2)$

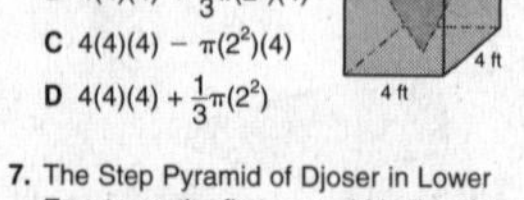

6. Approximately how many fluid ounces of water can the paper cup hold? (*Hint:* 1 fl oz ≈ 1.805 in^3)

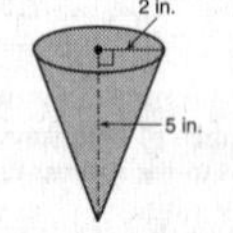

F 10.9 fl oz
Ⓖ 11.6 fl oz
H 32.7 fl oz
J 36.3 fl oz

7. The Step Pyramid of Djoser in Lower Egypt was the first pyramid in the history of architecture. Its original height was 204 feet, and it had a rectangular base measuring 411 feet by 358 feet. Which is the best estimate for the volume of the pyramid in cubic yards?

Ⓐ 370,570 yd^3
B 1,111,709 yd^3
C 3,335,128 yd^3
D 10,005,384 yd^3

Copyright © by Holt, Rinehart and Winston. All rights reserved.

LESSON 10-7
Reading Strategies
Use a Concept Map

Use the concept maps below to help you understand and use formulas for volume.

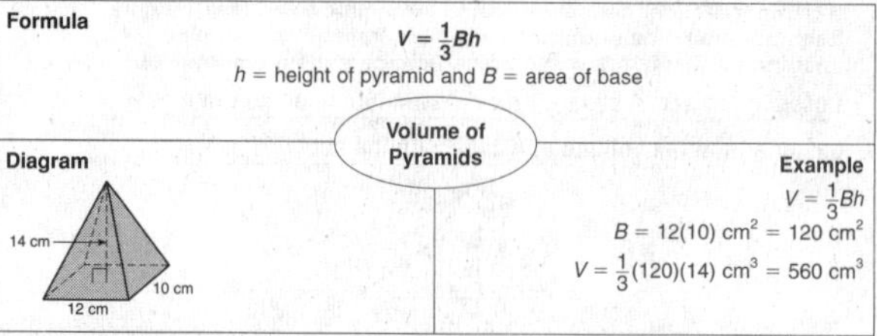

Formula
$V = \frac{1}{3}Bh$
h = height of pyramid and B = area of base
Volume of Pyramids
Diagram
14 cm
10 cm
12 cm
Example
$V = \frac{1}{3}Bh$
$B = 12(10) \text{ cm}^2 = 120 \text{ cm}^2$
$V = \frac{1}{3}(120)(14) \text{ cm}^3 = 560 \text{ cm}^3$

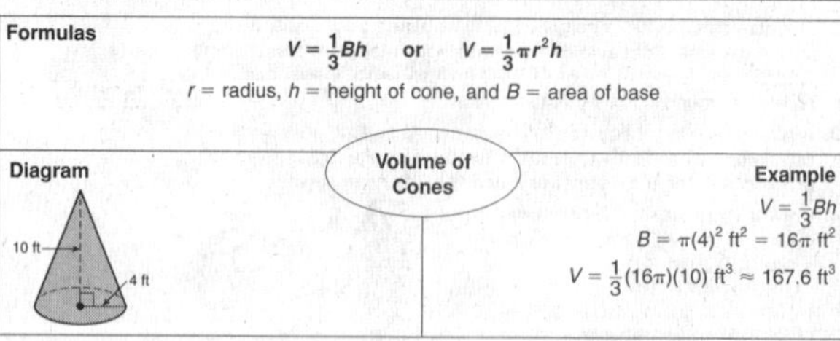

Formulas
$V = \frac{1}{3}Bh$ or $V = \frac{1}{3}\pi r^2 h$
r = radius, h = height of cone, and B = area of base
Volume of Cones
Diagram
10 ft
4 ft
Example
$V = \frac{1}{3}Bh$
$B = \pi(4)^2 \text{ ft}^2 = 16\pi \text{ ft}^2$
$V = \frac{1}{3}(16\pi)(10) \text{ ft}^3 \approx 167.6 \text{ ft}^3$

Find the volume of each figure. Round to the nearest tenth if necessary.

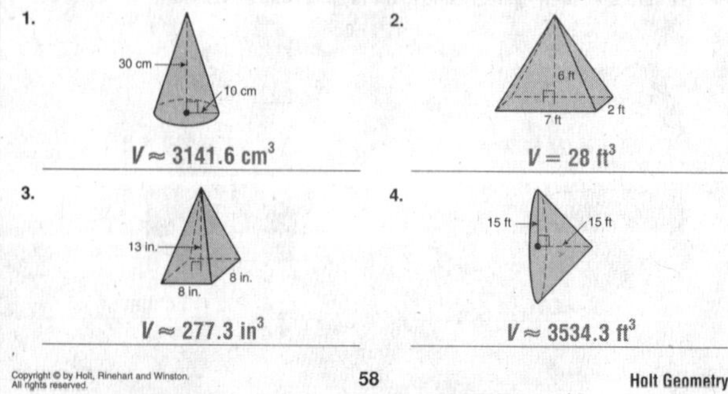

1. $V \approx 3141.6 \text{ cm}^3$
2. $V = 28 \text{ ft}^3$
3. $V \approx 277.3 \text{ in}^3$
4. $V \approx 3534.3 \text{ ft}^3$

Copyright © by Holt, Rinehart and Winston. All rights reserved.

Copyright © by Holt, Rinehart and Winston. All rights reserved.

LESSON 10-8

Practice A

Spheres

Write each formula.

1. volume of a sphere with radius *r* — $V = \frac{4}{3}\pi r^3$
2. surface area of a sphere with radius *r* — $S = 4\pi r^2$

Find each measurement. Give your answers in terms of π.

3. 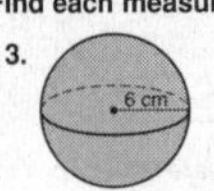

the volume of the sphere

$V = 288\pi \text{ cm}^3$

4. 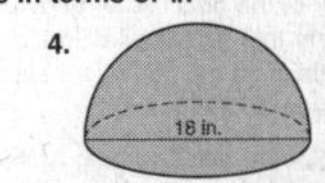

the volume of the hemisphere

$V = 486\pi \text{ in}^3$

5. the radius of a sphere with a volume of $36{,}000\pi \text{ mm}^3$ — $r = 30$ mm

6. Margot is thirsty after a 5-km run for charity. The organizers offer the containers of water shown in the figure. Margot wants the one with the greater volume of water. Tell which container Margot should pick.

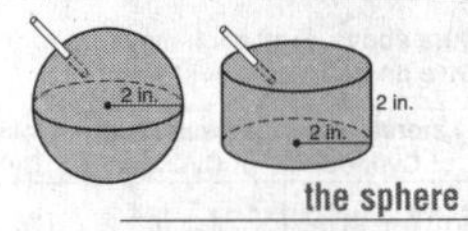

the sphere

Find the surface area of each sphere. Give your answers in terms of π.

7. 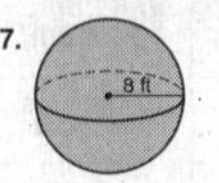

$S = 256\pi \text{ ft}^2$

8. the surface area of a sphere with volume $\frac{256\pi}{3}$ yd³

$S = 64\pi \text{ yd}^2$

Complete Exercises 9–11 to describe the effect on the volume and the surface area of multiplying the radius of a sphere by 3.

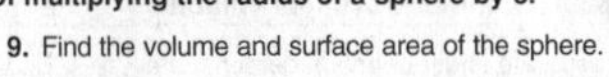

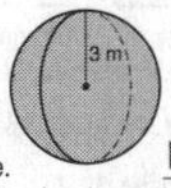

9. Find the volume and surface area of the sphere. — $V = 36\pi \text{ m}^3$; $S = 36\pi \text{ m}^2$
10. Find the volume and surface area of the sphere after the radius is multiplied by 3. — $V = 972\pi \text{ m}^3$; $S = 324\pi \text{ m}^2$
11. Describe the effect on the volume and surface area of multiplying the radius of the sphere by 3.

The volume is multiplied by 27. The surface area is multiplied by 9.

12. Find the volume and surface area of the composite figure. Give your answers in terms of π.

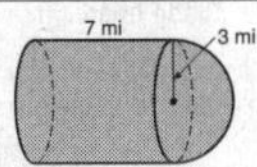

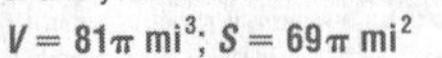

$V = 81\pi \text{ mi}^3$; $S = 69\pi \text{ mi}^2$

Copyright © by Holt, Rinehart and Winston. All rights reserved.

LESSON 10-8

Practice B

Spheres

Find each measurement. Give your answers in terms of π.

1. 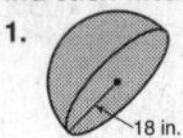

the volume of the hemisphere

$V = 3888\pi \text{ mm}^3$

2. 

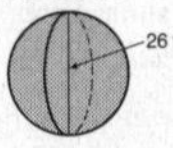

the volume of the sphere

$V = \frac{8788\pi}{3} \text{ ft}^3 = 2929\frac{1}{3}\pi \text{ ft}^3$

3. the diameter of a sphere with volume $\frac{500\pi}{3}$ m³ — $d = 10$ m

4. The figure shows a grapefruit half. The radius to the outside of the rind is 5 cm. The radius to the inside of the rind is 4 cm. The edible part of the grapefruit is divided into 12 equal sections. Find the volume of the half grapefruit and the volume of one edible section. Give your answers in terms of π.

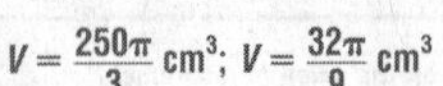

$V = \frac{250\pi}{3} \text{ cm}^3$; $V = \frac{32\pi}{9} \text{ cm}^3$

Find each measurement. Give your answers in terms of π.

5. 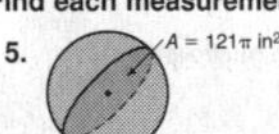

the surface area of the sphere

$S = 484\pi \text{ in}^2$

6. 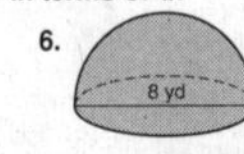

the surface area of the closed hemisphere and its circular base

$S = 48\pi \text{ yd}^2$; $S = 16\pi \text{ yd}^2$

7. the volume of a sphere with surface area 196π km² — $V = \frac{1372\pi}{3} \text{ km}^3 = 457\frac{1}{3}\pi \text{ km}^3$

Describe the effect of each change on the given measurement of the figure.

8. 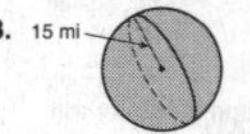

surface area
The dimensions are divided by 4.

The surface area is divided by 16.

9.  36 m

volume
The dimensions are multiplied by $\frac{2}{5}$.

The volume is multiplied by $\frac{8}{125}$.

Find the surface area and volume of each composite figure. Round to the nearest tenth.

10. 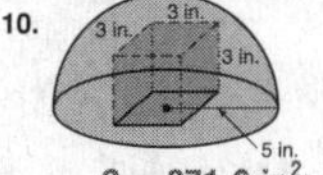

$S \approx 271.6 \text{ in}^2$; $V \approx 234.8 \text{ in}^3$

11. 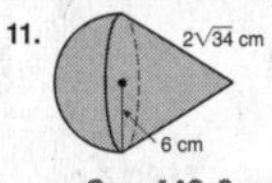

$S \approx 446.0 \text{ cm}^2$; $V \approx 829.4 \text{ cm}^3$

Copyright © by Holt, Rinehart and Winston. All rights reserved.

LESSON 10-8

Practice C

Spheres

1. A sphere has radius *r*. Draw a composite figure made up of a square prism (not a cube) and a square pyramid that has the same volume as the sphere.

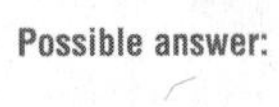

Possible answer:

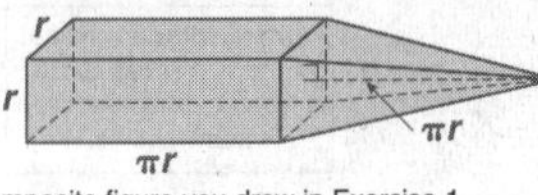

2. Find the surface area of the composite figure you drew in Exercise 1.

Possible answer: $S = 4\pi r^2 + r^2 + r^2\sqrt{4\pi^2 + 1}$

3. Consider a composite figure made up of a cylinder and a cone that has the same volume as a sphere with radius *r*. Find the figure's surface area.

$S = 3\pi r^2 + \pi r^2\sqrt{2}$

Use the figure for Exercises 4–6. The figure shows a hollow, sealed container with some water inside.

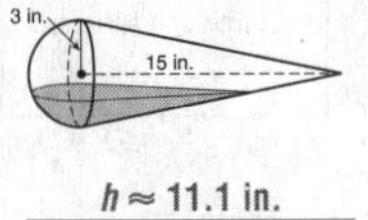

4. There is just enough water in the container to exactly fill the hemisphere. The container is held so that the point of the cone is down and the altitude of the cone is exactly vertical. Find the height of the water in the cone. Round to the nearest tenth. — $h \approx 11.1$ in.
5. Suppose the amount of water in the container is exactly enough to fill the cone. The container is held so that the hemisphere is down and the altitude of the cone is exactly vertical. Find the height of the water in the container. Round to the nearest tenth. — $h \approx 6.9$ in.
6. Find the height of the cone with the same radius if the container were made so that the water would exactly fill either the hemisphere or the cone. — $h = 6$ in.
7. A sphere has center (0, 0, 0). Its surface passes through the point (*x*, *y*, *z*). Find the sphere's surface area and volume.

$S = 4\pi(x^2 + y^2 + z^2)$; $V = \frac{4}{3}\pi(x^2 + y^2 + z^2)^{\frac{3}{2}}$

Use the figure for Exercises 8–10. The figure shows a can of three tennis balls. The can is just large enough so that the tennis balls will fit inside with the lid on. The diameter of each tennis ball is 2.5 in. Give exact fraction answers.

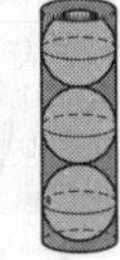

8. Find the total volume of the can. — $V = \frac{375\pi}{32} \text{ in}^3$
9. Find the volume of empty space inside the can. — $V = \frac{125\pi}{32} \text{ in}^3$
10. Tell what percent of the can is occupied by the tennis balls. — $66\frac{2}{3}\%$

Copyright © by Holt, Rinehart and Winston. All rights reserved.

LESSON 10-8

Reteach

Spheres

Volume and Surface Area of a Sphere		
Volume	The volume of a sphere with radius *r* is $V = \frac{4}{3}\pi r^3$.	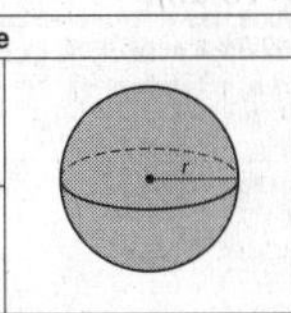
Surface Area	The surface area of a sphere with radius *r* is $S = 4\pi r^2$.	

Find each measurement. Give your answer in terms of π.

1. the volume of the sphere

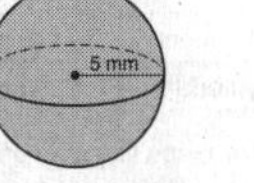

$V = \frac{500\pi}{3} \text{ mm}^3$

2. the volume of the sphere

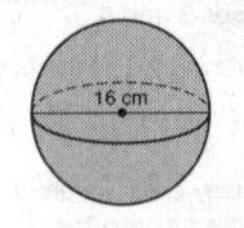

$V = \frac{2048\pi}{3} \text{ cm}^3$

3. the volume of the hemisphere

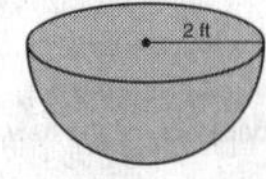

$V = \frac{16\pi}{3} \text{ ft}^3$

4. the radius of a sphere with volume 7776π in³

$r = 18$ in.

5. the surface area of the sphere

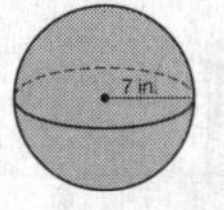

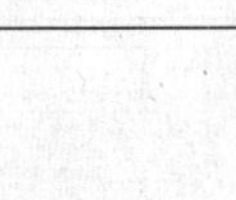

$S = 196\pi \text{ in}^2$

6. the surface area of the sphere

$S = 400\pi \text{ m}^2$

Copyright © by Holt, Rinehart and Winston. All rights reserved.

Copyright © by Holt, Rinehart and Winston. All rights reserved.

LESSON 10-8
Reteach
Spheres continued

The radius of the sphere is multiplied by $\frac{1}{4}$. Describe the effect on the surface area.

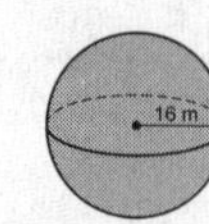

original surface area:		new surface area, radius multiplied by $\frac{1}{4}$:	
$S = 4\pi r^2$		$S = 4\pi r^2$	
$= 4\pi(16)^2$	$r = 16$	$= 4\pi(4)^2$	$r = 4$
$= 1024\pi\ m^2$	Simplify.	$= 64\pi\ m^2$	Simplify.

Notice that $1024 \cdot \frac{1}{16} = 64$. If the dimensions are multiplied by $\frac{1}{4}$, the surface area is multiplied by $\left(\frac{1}{4}\right)^2$, or $\frac{1}{16}$.

Describe the effect of each change on the given measurement of the figure.

7. surface area
The radius is multiplied by 4.

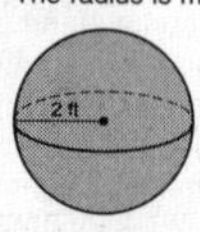

The surface area is multiplied by 16.

8. volume
The dimensions are multiplied by $\frac{1}{2}$.

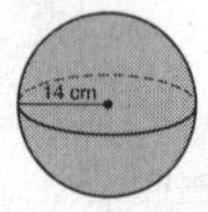

The volume is multiplied by $\frac{1}{8}$.

Find the surface area and volume of each composite figure. Round to the nearest tenth.

9. *Hint:* To find the surface area, add the lateral area of the cylinder, the area of one base, and the surface area of the hemisphere.

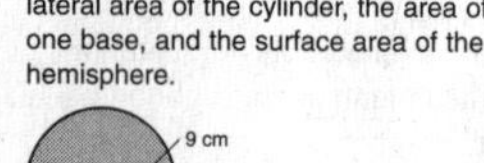
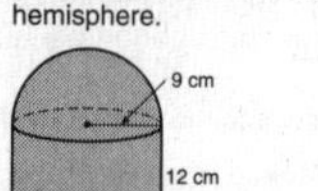

$S \approx 1442.0\ cm^2$; $V \approx 4580.4\ cm^3$

10. *Hint:* To find the volume, subtract the volume of the hemisphere from the volume of the cylinder.

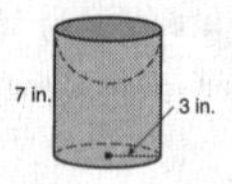

$S \approx 216.8\ in^2$; $V \approx 141.4\ in^3$

LESSON 10-8
Challenge
Spheres, Cylinders, and Archimedes

The Greek mathematician Archimedes (ca. 287–212 B.C.), a native of Syracuse, Sicily, is considered one of the greatest mathematicians of all time. He is perhaps best known for his contributions to the field of mechanics, such as the invention of the Archimedean screw and the discovery of the principle of buoyancy. However, it was geometry that Archimedes found most fascinating, and he established the first exact expressions for the volume and surface area of a sphere.

In a work titled *On the Sphere and Cylinder,* Archimedes examined the relationship between a right cylinder and a sphere inscribed in it. He considered this work to be so significant that he requested a representation of a cylinder and inscribed sphere to be engraved on his tombstone.

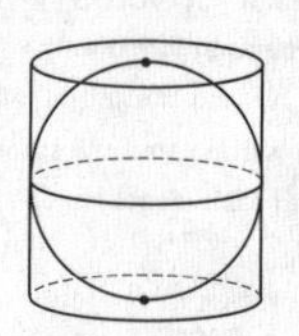

When a sphere is inscribed in a right cylinder, a diameter of the sphere lies on the lateral surface of the cylinder, and the sphere intersects the cylinder at the centers of the base.

Refer to the figure above. Find each measure for the given radius. Give answers in exact form.

	Radius	Lateral Area of Cylinder	Surface Area of Cylinder	Volume of Cylinder	Surface Area of Sphere	Volume of Sphere
1.	2 in.	$16\pi\ in^2$	$24\pi\ in^2$	$16\pi\ in^3$	$16\pi\ in^2$	$\frac{32}{3}\pi\ in^3$
2.	5 cm	$100\pi\ cm^2$	$150\pi\ cm^2$	$250\pi\ cm^3$	$100\pi\ cm^2$	$\frac{500}{3}\pi\ cm^3$
3.	10 ft	$400\pi\ ft^2$	$600\pi\ ft^2$	$2000\pi\ ft^3$	$400\pi\ ft^2$	$\frac{4000}{3}\pi\ ft^3$
4.	1.5 m	$9\pi\ m^2$	$13.5\pi\ m^2$	$6.75\pi\ m^3$	$9\pi\ m^2$	$\frac{9}{2}\pi\ m^3$

5. Refer to the table in Exercises 1–4. Archimedes discovered that two of the sets of measures related the sphere to the cylinder by the ratio 2 : 3. Which measures are they?

sphere surface to cylinder surface; sphere volume to cylinder volume

6. Refer to your answer to Exercise 5. On a separate sheet of paper, demonstrate algebraically why the two ratios you identified are always 2 : 3.

Check students' work.

7. In his work *Measurement of a Circle,* Archimedes demonstrated that the value of π was between $\frac{223}{71}$ and $\frac{22}{7}$. That is, he approximated the value of π correctly to the nearest hundredth, or 3.14.

Using your library or the Internet as a resource, research the method that Archimedes used to make his approximation. Write your results on a separate sheet of paper.

LESSON 10-8
Problem Solving
Spheres

1. A globe has a volume of $288\pi\ in^3$. What is the surface area of the globe? Give your answer in terms of π.

$144\pi\ in^2$

2. Eight bocce balls are in a box 18 inches long, 9 inches wide, and 4.5 inches deep. If each ball has a diameter of 4.5 inches, what is the volume of the space around the balls? Round to the nearest tenth.

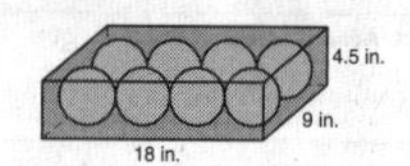

$347.3\ in^3$

Use the table for Exercises 3 and 4.

Ganymede, one of Jupiter's moons, is the largest moon in the solar system.

Moon	Diameter
Earth's moon	2160 mi
Ganymede	3280 mi

3. Approximately how many times as great as the volume of Earth's moon is the volume of Ganymede?

about 3.5 times

4. Approximately how many times as great is the surface area of Ganymede than the surface area of Earth's moon?

about 2.3 times

Choose the best answer.

5. What is the volume of a sphere with a great circle that has an area of $225\pi\ cm^2$?

A $300\pi\ cm^3$ C $2500\pi\ cm^3$
B $900\pi\ cm^3$ (D) $4500\pi\ cm^3$

6. A hemisphere has a surface area of $972\pi\ cm^2$. If the radius is multiplied by $\frac{1}{3}$, what will be the surface area of the new hemisphere?

F $36\pi\ cm^2$ H $162\pi\ cm^2$
(G) $108\pi\ cm^2$ J $324\pi\ cm^2$

7. Which expression represents the volume of the composite figure formed by the hemisphere and cone?

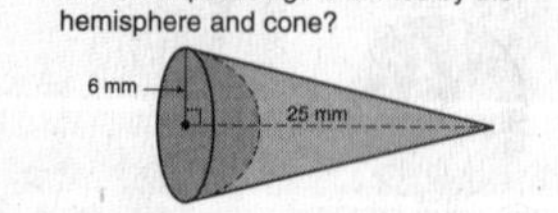

A $52\pi\ mm^3$ C $276\pi\ mm^3$
(B) $156\pi\ mm^3$ D $288\pi\ mm^3$

8. Which best represents the surface area of the composite figure?

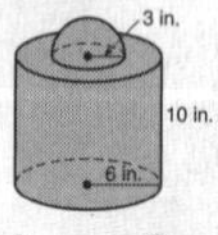

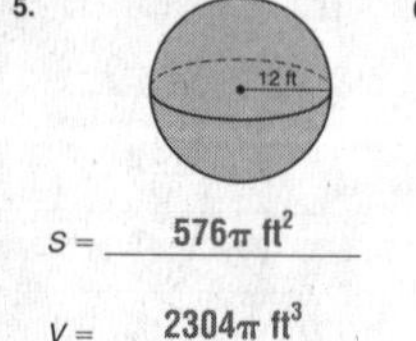

F $129\pi\ in^2$ (H) $201\pi\ in^2$
G $138\pi\ in^2$ J $210\pi\ in^2$

LESSON 10-8
Reading Strategies
Focus on Vocabulary

The diagram below describes the parts of a sphere and gives you the formulas for surface area and volume of a sphere.

A **hemisphere** is half a sphere. A **great circle** divides a sphere into two hemispheres.

A **sphere** is the locus of points in space that are a fixed distance from the **center** of the sphere.

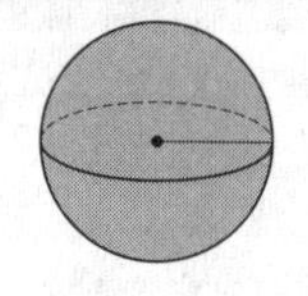

A **radius** r connects the center of the sphere to any point on the sphere.

The formula for the **volume** of a sphere is $V = \frac{4}{3}\pi r^3$.

The formula for the **surface area** of a sphere is $S = 4\pi r^2$.

Answer the following.

1. The radius of a sphere connects the center of the sphere to any point on the sphere.
2. A(n) hemisphere is half a sphere.
3. The radius of a sphere is 6 centimters. What is the radius of its great circle? 6 cm
4. The volume of a sphere is $900\pi\ ft^3$. What is the volume of one of its hemispheres? $450\pi\ ft^3$

Find the surface area and volume of each sphere. Give your answers in terms of π.

5. (12 ft)
$S = 576\pi\ ft^2$
$V = 2304\pi\ ft^3$

6. (22 in.)
$S = 484\pi\ in^2$
$V = \frac{5324\pi}{3}\ in^3$

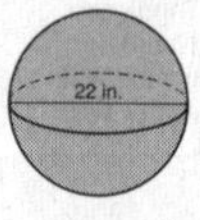

7.

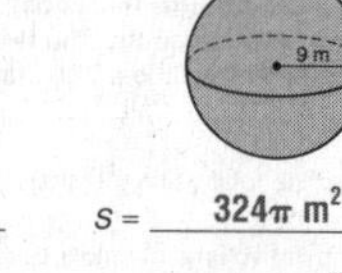

$S = 324\pi\ m^2$
$V = 972\pi\ m^3$

Copyright © by Holt, Rinehart and Winston. All rights reserved.